AF805103

UNIVERSITY OF WESTMINSTER

Failure to return or renew overdue books may result in suspension of borrowing rights at all University of Westminster libraries.

Due for return on:

28 NOV 1995
17 MAR 1997
21. FEB 96
12. MAR 97
20 MAY 1996
-6 DEC 1999
13. JUN 96
23. SEP 96
-4 NOV 1996
-4 NOV 1996
04. MAR 97

Information Resource Services **Riding House Street Library**
37–49 Riding House Street London W1P 7PT
Telephone 071 911 5000 ext 3887
0056

22 0274805 0

The Royal Institute of International Affairs is an independent body which promotes the rigorous study of international questions and does not express opinions of its own. The opinions expressed in this publication are the responsibility of the authors.

2202748050

327.5204
DEF

THE SECURITY CHALLENGES FOR JAPAN AND EUROPE IN A POST-COLD WAR WORLD

VOLUME 3

THE DEFENCE TRADE: DEMAND, SUPPLY AND CONTROL

edited by

TREVOR TAYLOR
AND RYUKICHI IMAI

ROYAL INSTITUTE OF INTERNATIONAL AFFAIRS
and
INSTITUTE FOR INTERNATIONAL POLICY STUDIES

We are grateful to the Sasakawa Foundation in Tokyo, which provided the principal funding for this joint project.

© Royal Institute of International Affairs, 1994

First published in Great Britain in 1994 by
Royal Institute of International Affairs
Chatham House, 10 St James's Square, London SW1Y 4LE

Distributed worldwide by The Brookings Institution, 1775 Massachusetts Avenue, Northwest, Washington, DC 20036-2188, USA

All rights reserved. No part of this publication may be reproduced, stored in a retrieval system, or transmitted by any other means without the prior written permission of the copyright holder. Please direct all inquiries to the publishers.

British Library Cataloguing in Publication Data

A CIP catalogue record for this book is available from the British Library.

ISBN 0 905031 74 1

Text designed and set by Hannah Doe
Cover design by Youngs Design in Production
Printed and bound in Great Britain by Redwood Books

CONTENTS

ACKNOWLEDGMENTS

This multi-author volume, the third in a series, has been made possible by the dedication of a range of people. Many staff at the IIPS worked hard with regard both to the preparation and operation of the initial conference in June 1993, and to the follow-up work afterwards. Particular appreciation is due to Takashi Kawakami, who led administrative support in Tokyo, and Katherine Markulin, Publications Manager at IIPS. Emma Matanle of the International Security Programme at the RIIA managed administrative matters in London and spent numerous hours preparing the text for publication. Many thanks are due to Margaret May, head of the Publications Department at Chatham House, to Rosamund Howe for her editing, and to Hannah Doe for text design and production. We are grateful to the Sasakawa Foundation in Tokyo, which provided the principal funding for this joint project, and to British Aerospace plc, which also provided helpful support. As ever, any shortcomings in the work must lie with the editors in London and Tokyo.

November 1993

Trevor Taylor
Ryukichi Imai

CONTRIBUTORS

Yves Boyer, Deputy Director, CREST Ecole Polytechnique
Charles Drace-Francis, Government Affairs Director, British Aerospace
Peter van Ham, Research Fellow, Institute of Security Studies, WEU
Ryukichi Imai, Distinguished Scholar, Institute for International Policy Studies
Takeshi Ito, Deputy Director, Technology Research Division, MITI
Mitsuru Kurosawa, Professor, Faculty of Law, Osaka University
Colonel Yoshiaki Nakagawa, Senior Research Fellow, Institute for International Policy Studies
Martin Navias, Lecturer, King's College, London University
Trevor Taylor, Professor of International Relations, University of Staffordshire; Associate Fellow, International Security Programme, Royal Institute of International Affairs
Osamu Watanabe, Director General, MITI

ABBREVIATIONS

ACDA	US Arms Control and Disarmament Agency
APEC	Asia-Pacific Economic Conference
ASEAN	Association of South-East Asian Nations
ATBM	Anti-Tactical Ballistic Missile
AWAC	Airborne Warning and Control System
BAe	British Aerospace Plc
BTWC	Convention on the Prohibition of the Development, Production and Stockpiling of Bacteriological (Biological) and Toxin Weapons and on their Destruction
CBM	Confidence-Building Measure
CD	Conference on Disarmament
CFE	Conventional Forces in Europe
CNN	Cable News Network
COCOM	Coordinating Committee (on Multilateral Export Controls)
CRS	US Congressional Research Service
CTB	Comprehensive Test Ban
CWC	Chemical Weapons Convention
FSU	Former Soviet Union
G7	Group of Seven (Canada, France, Germany, Italy, Japan, UK, US)
HEU	Highly Enriched Uranium
IAEA	International Atomic Energy Agency
ICBM	Intercontinental Ballistic Missile
INF	Intermediate-Range Nuclear Forces
ISTC	International Science and Technology Center

MBFR	Mutual and Balanced Force Reduction
MIRV	Multiple Independently Targetable Re-Entry Vehicle
MITI	Ministry for International Trade and Industry
MLRS	Multiple Launch Rocket System
MTCR	Missile Technology Control Regime
NNWS	Non-Nuclear Weapons State
NPT	Nuclear Non-Proliferation Treaty
NTM	National Technical Means
NWFZ	Nuclear Weapon-Free Zone
NWS	Nuclear Weapons State
ODA	Official Development Aid
OPCW	Organization for the Prohibition of Chemical Weapons
SIOP	Single Integrated Operating Plan
SIPRI	Stockholm International Peace Research Institute
SLBM	Submarine-Launched Ballistic Missile
SRAM	Short-Range Attack Missile
SS	Surface-to-Surface (Soviet Missile)
SSC	Superconducting Super Corridors
START	Strategic Arms Reduction Treaty
UNSCOM	United Nations Special Commission on Iraq

INTRODUCTION

TREVOR TAYLOR

A significant consequence of the conflict over Kuwait, followed by the alarm generated by North Korea's efforts in the nuclear and missile field, is that the proliferation of weapons of mass destruction and the trade in conventional arms have been seen as increasingly important issues in international affairs. There is wide agreement in the developed world with the assessment of the US Secretary of State, Warren Christopher, that 'We face no more urgent security threat than the potential spread of weapons of mass destruction and the means of delivering them'. This volume reflects the concern with these subjects in both Japan and Western Europe.

Both this volume and the conference which generated it reflect two fundamental judgments. The first is that the area of export control is a difficult one for governments in that the actual implementation of almost any policy guidelines can raise a plethora of practical problems. Governments have to develop and update lists of products of specific concern and they have to communicate these lists to their industry. Procedures for issuing licences have to be established which do not place an unacceptable burden on industry to complete forms and provide information, which bring forth considered and authoritative judgments from a range of government departments, and which generate the issuing of licences without the undue delay which may cost a company business. It is an aspect of government policy where the devil is clearly in the detail and some attention is devoted to these issues by authors here.

The second judgment is that the trade in defence-related technology can be seen as comprising four different, if somewhat overlapping, sectors. The first two comprise the transfers of conventional arms and non-conventional arms themselves, the latter being defined as chemical, biological and nuclear weapons and the ballistic missiles which could deliver them over significant distances. There is of course a major international effort to outlaw the trade in non-conventional arms completely. The third and fourth sectors comprise transfers of dual-use technologies which can contribute to the development and manufacture of conventional and non-conventional arms to countries which are seeking to make their own systems.

While these categories can be treated individually for some purposes, it is often necessary to address the composite picture of the trade in defence equipment and defence-related technology. Not least this is because a state denied one sort of technology may well seek another form of defence technology in recompense. It is quite clear that states which have been, or have reason to fear they may be, subjected to an arms embargo devote considerable effort to building up their own defence industrial capability.

The chapters in this book are broad ranging and reflect industrial as well as governmental perspectives. The book begins with Ryukichi Imai locating export control problems in the wider context of science and technology issues on the international agenda at the end of the Cold War.

Attention is then paid to the demand for non-conventional arms and how proliferation can be prevented or at least delayed. Yoshiaki Nakagawa considers the demand for nuclear weapons, noting both the perceived prestige value of such systems and their utility in political conflict, while Martin Navias produces an alarming catalogue of the many programmes concerned with ballistic missiles, chemical weapons and biological weapons development. Moving to the control side of the equation, Mitsuru Kurosawa analyses the prospects for the nuclear non-proliferation regime, while Peter van Ham's analysis includes the Missile Technology Control Regime, the Chemical Weapons Convention and the Biological Weapons Convention as elements of non-proliferation regimes.

The incentives to export non-conventional arms and associated technology are modest compared with the disadvantages, but in the conventional arms trade, the pressures to export on companies and their supervising governments are significant and Trevor Taylor's analysis looks at the factors behind both the export and import of arms. The economic and strategic pressures for arms exports at a time of diminishing defence budgets at home are quite apparent. Takeshi Ito considerers controls on conventional arms exports, and includes analysis of the UN Register of Conventional Arms, which was set up as a joint EC–Japan initiative in the General Assembly. Japan's policy, of course, at considerable cost to its own defence budget, is not to export arms at all (except to its close ally, the United States).

Yves Boyer's contribution points to some of the new difficulties emerging in this broad area, in that it suggests that the distinctions between conventional and non-conventional weapons are being eroded by technological change, and it points to areas such as non-lethal weapons which could emerge as future foci for arms control.

This book concludes with a contribution from Charles Drace-Francis, a diplomat currently working in industry, which addresses issues of controls in practice in the past and present. This chapter, coupled with the practical control issues which Ambassador Imai mentions in his chapter, stresses our concern with policy in practice as well as principle. With these latter concerns in mind, the book includes two useful appendices. Appendix 1 contains the text of an address to the IIPS–RIIA conference by MITI Director General Osamu Watanabe in which he discusses Japanese policy on the export of dual-use technology. Appendix 2 is the text of a report for MITI from a group chaired by Ambassador Imai on the future of Japanese export controls. The findings of the report have been circulated by the government and appear likely to become part of Japanese policy.

Japan is a very minor exporter in the arms trade, with the US being its only customer; but it is a major arms importer and an exporter of high technology, much of which has a dual-use character. Western Europe is prominent in the world arms market although selling in dollar terms much less than the United States, and the Kuwait crisis and its aftermath exposed just how much Western Europe had contributed to the Iraqi defence effort. Despite their rather different starting points, for both Western Europe and Japan controlling the export of defence-related technology will be increasingly problematic. This volume represents an effort at consultation which it is hoped will encourage Japanese–European cooperation on the adoption of workable, equitable and effective multilateral controls.

CHAPTER 1

CONTROLLING DEFENCE-RELATED TECHNOLOGY AFTER THE COLD WAR: THE POLITICAL–TECHNOLOGICAL ENVIRONMENT

RYUKICHI IMAI

Controls on defence technology transfers are not shaped and operated in a political vacuum, but are much influenced by wider developments both in the international system as a whole and in particular regions. Moreover, technology transfer controls are but one of several interrelated technologically based problems raised by the end of the Cold War. These include how to fund and direct future research and development efforts, how to use technology to promote economic development in poorer and richer states, how to sustain desired defence industrial capabilities and how to manage the relationship of technology with the natural environment. The end of the Cold War has brought about many changes, and it may be useful to enumerate some of the most prominent, and to speculate as to where they may lead the world.

Changing ground rules in world politics

First, the significance of nuclear deterrence as the backbone of the East–West confrontation has changed. One no longer hears about flexible response, mutually assured destruction, or non-nuclear offensive defence in Europe. Practically all activities in nuclear warhead development in the United States and the former Soviet Union (FSU) have been suspended, and negotiations on a comprehensive underground nuclear test ban may resume. This could mean that the nine safety-related tests (connected with design changes stemming from the adoption of insensitive high explosives) now planned for execution in Nevada by 1996 would be the last among the former nuclear superpowers.

The United States and the former Soviet Union have agreed that they no longer need 12,000 strategic warheads each to defend their security, or even 6,000 (as agreed in START I), but that as few as 3,000 (START II) are sufficient. These START II numbers are singularly unconvincing and devoid of any logical and theoretical justifications, so that:

(a) there is no credible revised Single Integrated Operating Plan (SIOP) to justify warhead numbers, to determine whether land-based remaining ICBM silos should be included as targets for a first strike, etc;
(b) the international balance of power, which should take into account China, France and the UK, has not been discussed. Deterrence with numbers in the region of 3,000 warheads requires more careful planning and more elaborate attention to the worldwide nuclear deterrence system than it did when the sheer numerical superiority of the US and the Soviet Union was obvious;
(c) no credible and convincing scenario for ratification and implementation of these treaties can be seen on the horizon. The objection raised by Ukraine to the Lisbon Protocol involves something other than counteraction to Russian domination. On START II, the other republics were not even consulted. Ukraine seems ever more drawn to developing and holding onto the status of a nuclear weapons state and is particularly reluctant to see the 46 SS24s on its territory removed. Nuclear weapons on Ukrainian territory are more likely to become non-operational because of Russia's failure to maintain them than through arms control agreement implementation.

The world has just started the process of discovering and defining a situation in which nuclear deterrence is no longer the sole means of security. Yet, large arsenals of equally destructive tactical weapons remain almost untouched. It is wrong to say that the nuclear arsenals of the great powers have no role to play. They simply have a significance that is less well defined, less balanced, and more delicate, thus making any game plan more complicated.

Second, regional conflicts, including those that are large-scale and complex, such as in the Gulf, are becoming more unmanageable. For a long time the United States was prepared to protect the oil flow from the Middle East to the West against any possible intervention by the Soviet Union.[1] Now that it is clear that FSU military forces would not (at least for the moment) think of invading the region, the significance of the Arab/Israeli conflict has returned to a more normal status, that of a regional conflict which does not inevitably involve great-power confrontation. Yet the Middle East, as a region containing two-thirds of the world's known oil resources, will continue to attract outside attention, especially because of fears of fundamentalist Islamic rule. Cambodia is an important test case for a large-scale UN intervention involving both nation-building and nation-governing, but it remains essentially a regional problem with little possibility of igniting a larger conflict involving neighbouring powers. Latin America too has always been fairly isolated, hence the Security Council's refusal to recognize that there is a major economic, humanitarian and stability issue in Haiti, about which even the Organization of American States has tried to take an uncharacteristic initiative. Bosnia/Herzegovina, except for the fact that the area is both physically and mentally closer to the European heart than Somalia, Mozambique or Haiti, could also be termed a basically regional problem if only CNN could be persuaded to give equal coverage to the other trouble spots around the globe.

The world is full of regional conflicts, and each does not have to bring in major powers. In other words, various levels of regional conflict may be expected and tolerable. Certainly, the United Nations should not try to get involved in every conflict or expect that somebody will cover all the costs (human and material) of large blue-helmet operations.

Third, rather crude weapons of mass destruction could become significant threats. Such weapons could involve nuclear fission detonation capabilities, nerve gas, or biological weapons. The smallest-sized publicly known nuclear warhead is the US army's 155 mm W48, which has a height of about 85 cm, weighs about 54 kg, and has a yield between 0.1 and 1 kt (TNT equivalent).[2] Although difficult to manufacture and make operational, these warhead dimensions are enough to support many spy-story versions of the 'bomb in the city-centre' syndrome, which takes warheads from clean, artificial, sophisticated war-game computer rooms to more unmanageable and dirty field situations.

Changing ground rules in defence economics and industry

The world spent one trillion dollars on military expenditures in 1990 (assuming that worldwide gross products can be calculated and that military spending represents about 4 to 5 per cent of total GNP). Weapons production hovered at around $300 billion, of which about $50 billion went into the international weapons trade, and close to $40 billion involved the developing world.[3] Yet there may be serious and significant change under way in the traditional pattern of military spending, arms procurement and arms transfers.

Actual and possible cancellation of US arms projects such as the W88 and W89 warheads, the Trident D-5 SLBMs, new SRAM missiles, B-2 stealth and supersonic bomber, and Midgetman mobile single-warhead ICBMs means significant restructuring for the US defence industry. Arms production in the former Soviet Union has also changed so that no more T-80s, Blackjacks, SS-25s or Suhoy-27s are required. With the possible implementation of START II and subsequent nuclear disarmament, high-tech centres such as Los Alamos, Lawrence Livermore, Arzamus-16 and Cheryabinsk-70 may conceivably become inactive, leaving several thousand capable weapons scientists out of work.[4] Not only will that mean loss of employment, of production, and of R&D incentives, it may also lead to major changes in the international weapons trade, as manufacture adjusts to compensate for shrinking domestic markets. Countries may propose non-proliferation regimes as far as weapons of mass destruction are concerned, but they may also be more than willing to expand their export business in 'conventional' hardware capable of acting as a launch platform or delivery vehicle for systems capable of mass destruction. The recent expansion of the conventional weapons market into Saudi Arabia, Iran, China and Taiwan from countries including the US, Russia, China and even North Korea seems to indicate that jobs have to be protected, while so-called peace dividends are not easy to come by. The difficulties in converting military-oriented industry to civilian purposes have been described in detail by many writers,[5] who point to a number of reasons why the basic structure of defence-oriented technology and industry may be different from that of the civilian sector. Defence R&D is primarily quality-oriented with less concern for cost-effectiveness, while weapons producers normally have been provided with a captive purchaser offering a cost-plus-profit contract. The problems are basically the same in the United States and the FSU, although the latter has the further handicap of completely lacking marketing experience. Recent 50 per cent cuts in basic research expenditures by the computer giant IBM may indicate the type and the extent of changes research communities of the world have to face.

One additional consideration may be highlighted by taking Japan as an example. In spite of economic problems, defence expenditures in Japan have been rising in real value, reflecting (what was originally considered a major restraint) the budget ceiling of 1 per cent of GNP. Since Japan is basically not a weapons exporter, unit costs of most home-produced weapons are very high. For example, four licence-produced Boeing 767-based AWACs will cost 57 billion yen (US$500 million), which is almost double the original estimate for the planes. At the same time, Japan is probably one of the world's largest weapons importers, especially when licence purchase and production royalties are included. The US ACDA data (see Figure 1) indicate that the United States and Japan are among the major weapons importers of the developed world. Japan has decided to slash 580 billion yen from planned spending in a major mid-term defence reorganization programme (1991) originally budgeted at 22,750 billion yen (of the cut, 560 billion will be procurement reduction). To what extent Japan will be able to maintain even a semblance of an independent defence industry is open to question (see Tables 1.1, 1.2, 1.3 and 1.4 at end of chapter).

While a national debate may have started in Japan over increases in contributions to UN peacekeeping operations and the claim to a permanent seat in the Security Council, no one has jumped into the dangerous waters of combining such security obligations with the future of defence procurement.

Japanese experiences in technology acquisition

If I may become personal at this point, I would like to relate some experiences I had in the early 1980s while serving in the oil-rich Arab region. Many of my colleagues there asked me for the secret of Japan's rapid industrialization not only at the time of the Meiji Restoration but also after the defeat in the Second World War. Since money is, and was, not really the problem in the Middle East, my questioner thought that a knowledge of Japanese experiences could be of help in converting oil wealth into something more tangible, namely true industrialization. I tried to bring in economists, political scientists and industrialists to explain what happened in Japan.

Mid-nineteenth-century Japan, although still a feudal society, had already developed a considerable social infrastructure, including such things as primary schools, a currency exchange market and some industrial development (especially based on steel-making, including steel for Japanese swords). Japan opened its borders at a time when Western civilization had just entered the initial stage of industrialization. The country was determined to learn, and to that end hired many well-known scholars and experts at salaries higher than the top civil servants were paid. These 'hired' *gaijin* were treated with the utmost reverence, but since they were so expensive to keep, Japan tried to replace them with its own nationals as soon as possible. The point I was trying to make to my Arab friends was that in addition to effort, one has to have luck and timing, as Japan did when it began its own industrialization. Mirage fighters may be bought, but luck you cannot arrange for.

After Japan's defeat in the Second World War, had the Korean war not happened when it did, the United States would probably have continued to treat Japan as a small-scale agrarian society rather than as a military base with the capabilities to support major military operations. Later in the 1960s, the United States promoted free trade under the Bretton Woods accords and that country was so prosperous and so advanced that modern technology and accompanying know-how, including quality-control philosophy, were sold abroad at only the immediate cost of development. Later, people realized that projects that do not come to fruition in a nation's R&D efforts have to be included in the total national R&D costs, and that, consequently, higher price tags must be attached to individual technology transfers. Japan could not be blamed for taking advantage of US generosity in this regard because it itself was not aware of the tremendous break it was being given. (This historical consideration alone is a good reason why Japan should not judge its relations with the United States solely on the current trade and/or economic partnership.)

Table 1.5 indicates that, in spite of all the problems mentioned, the United States spends more on R&D than the rest of the advanced world put together. While there is no doubt that it will continue to be the technological leader of the world, a considerable portion of its R&D effort has traditionally gone to defence, producing high-quality but perhaps not so commercially competitive technology. All this is changing. The US government may or may not adopt official industrial policy and trade policy and it may or may not create a Department of International Trade and Industry (DITI) as some advocate, but there is no doubt that in addition to basic research, it is determined to promote more vigorously many types of process and manufacturing technology. It is still not clear, with the former head of the Congressional Office of Technology Assessment being appointed as the President's science adviser, whether the country will continue with expensive projects such as the Space Station and Superconducting Super Corridors (SSC). It would make most sense to consolidate international cooperative projects in these areas, and to establish coordinated development policies and projects in high-energy physics, space exploration, energy technology and biotechnology, and in environmental and other life sciences. Those who are 'lucky', those who have the 'necessary infrastructure', and those who can 'afford it' should get together.

It may also be possible to develop an international satellite communications network for disarmament verification, and for the monitoring of global environmental conditions. In order legitimately to assert restraints on technology transfers of certain kinds, developed countries should be able to explain what they are doing to promote technology for the future.

Controls on international technology transfer

A notable case of COCOM violation in which Japan was deeply involved was the Toshiba-Kongsberg incident of 1987.[6] Toshiba Machine Company, a subsidiary of Toshiba Corporation, together with a Norwegian firm, Kongsberg Vaapenfabrik, was selling multi-axis computer-controlled machine tools with accompanying software to the FSU, technology that would allegedly have made it possible for Soviet submarines to become quieter. The incident provided opportunities for the airing of sentiments against Japanese trade practices in general. The fact that the Norwegians reacted swiftly and immediately closed down the firm in question, while Japan moved slowly, gave rise to international criticism that finally forced the president of Toshiba Corporation to resign.

COCOM, ever since its creation in 1946, has been something of an enigma in that, while its headquarters were known to exist within the US embassy in Paris, much of the contents of the agreement were kept secret. Furthermore, as could be expected, investigations of COCOM cases, much like incidents involving drugs and spying, were usually the result of inside information. The handling of COCOM-related cases was difficult in Japan in particular because a 1969 court decision related to a trade exhibition in Beijing virtually ruled that COCOM-type restraints were against the basic spirit of Japanese foreign trade law (which stands for the principle of a free market).[7] Although the Japanese government tightened controls over sensitive technology exports through regulations (the Boeki Kanri Rei, or Trade Control Regulations, which come under the Foreign Trade Law), problems remain in the new technology transfer controls under the London Guidelines (for nuclear weapons-related technology and material), the Australia Group agreement concerning chemical weapons-related areas, and the Missile Technology Control Regime (MTCR). A somewhat similar situation exists with regard to defence technology in general. During the

1970s, when uranium enrichment technology became one of the central issues in future energy security, a problem developed in terms of Japanese legal structures to protect confidential information. It was realized that the country does not have laws to protect nationally classified information (except that related to the US–Japan Security Agreement), nor is there provision for secret patent. Whether enrichment technology is imported into Japan or whether it is domestically developed, if patent protection is applied for, the technology has to be made public. For technology not under patent protection, only civil damage suits (and civil servants' obligations to maintain secrecy about knowledge acquired through public service) can provide credible barriers against unintended dissemination of sensitive information. So far, this situation has not caused sufficient complications to make technology transfer or development especially difficult. Action has been taken to make firms concerned responsible for preventing information leakage, and the steps taken in the Toshiba-COCOM case inadvertently proved that such a system has a very significant restraining effect in Japan.

Japan plans to strengthen even further controls over transfers of technology that may be related to weapons of mass destruction. Most likely, the new system will compel potential exporters who 'know that possible diversion may be involved' to seek explicit export licences.

This concept, labelled as a 'know' regulation or 'catch-all' system, requires that a sufficient and systematic network of international information be available to ensure that knowledge may be shared among the countries concerned. In the absence of such a network of information, those without adequate advance protection may end up withdrawing from international trade, and that would be a serious development.

On the other hand, even with an international network, there is the possibility that an otherwise harmless common technology might one day turn out to be an important component of a weapon of mass destruction. One would also like to be careful and make sure that the exclusive club of knowledgeable exporters is not expanded too much and does not by chance include a possible importer/diverter who might abuse available information. There is certainly a need to coordinate this aspect of technology transfer control with restraint over foreign sales of conventional (especially high-tech) weapons since governments denied conventional weapons are likely to seek out other sources of defence technology.

The following seem to be issues that cut across the entire spectrum of export control:

(a) Can there be adequate, internationally agreed, common criteria to make particular technologies or equipments 'diversion suspect' as far as nuclear, chemical, or biological weapons and/or their delivery through ballistic missiles are concerned?
(b) Is it possible to name openly countries seen as suspect under the regime(s) specified above, and if so, how is their suspect nature to be established? COCOM was a notable example using open naming, as was the UN with regard to South Africa. Iraq's case required Security Council Resolution 687. What are the possible common criteria involved in these cases?
(c) If an export has to be suspended because of suspicion, the judgment (which may never be proven) will cause commercial loss to the firm involved. While containment was the guiding philosophy of East–West relations, national security and alliance obligations seemed to provide adequate grounds for denial of permission to export. What could be the new guiding principle in the post-Cold War period?
(d) What processes should be established for revising the list? What level of technological sophistication should be assumed on the part of possible diverters? Should there be a combined international body bringing together different regimes, which would meet regularly and agree on list updating? Should that require consensus or some sort of majority?

New aspects of the North–South relationship

Initially, the NPT safeguards were considered easy to implement because (primarily) only two chemical elements, uranium and plutonium, both of which are radioactive, toxic, and of little practical use outside the nuclear industry, were being dealt with. As the nations went through the process of implementing Article III of the NPT, they realized that the structure and content of the work, namely maintaining a sufficiently alert level of nuclear material accountancy and control in the peaceful fuel cycle, and issuing timely warnings against possible military diversion, involve a bureaucracy on the scale of the International Atomic Energy Agency.

Implementation of the Chemical Weapons Convention will be a much greater job, partly because the number of chemical elements involved is very large, ranging from hydrogen and oxygen to carbon, chloride, phosphate, etc. At the moment, those who are preparing the secretariat structure are dealing primarily with the declaration and destruction of chemical weapons already in the possession of signatories. Even that provides enough difficulties as more than twenty years of negotiations at the Geneva Conference on Disarmament have abundantly revealed. A look at the list of chemicals that have to be dealt with will amaze those with little background in chemistry. There are already problems with those chemicals that may have been left over from earlier (and thus unrecorded) chemical weapons activities, which may be discovered after nations have made their declarations. It is always possible that some country might discover a new chemical weapon (binary, maybe) that has not been listed in the Appendix. Challenge inspection provision is primarily aimed at undeclared chemical weapons that may be in existence at the time the treaty comes into force. To carry out inspections over literally millions of chemical industrial facilities throughout the world and to certify and verify that undeclared, illegal, clandestine activities in preparation for some future chemical warfare are not taking place is an indescribable task, which the secretariat now being prepared in The Hague would not even want to think about.

In both chemical and nuclear cases, the ability to control is based on the assumption that a reasonable industrial infrastructure is in existence in the countries to be scrutinized. In the case of the NPT safeguards, one more or less assumes that a national system of nuclear material accountancy is in effect, and that the international inspectorate does not have to start counting each atom of uranium and/or plutonium anew. The same applies for the chemical industry in that the central national authority must have a good grasp of individual chemical industrial firms, if only from the point of view of industrial safety or of taxation. If cooperation cannot be expected in such areas, then international control, verification, etc, become extremely difficult. One only needs to look into many of the documents made publicly available through such US government organizations as the Congressional Office of Technology Assessment or the National Academy of Sciences to find out that verification of nuclear arms control and disarmament agreements has to be based on availability of information from the other side, in addition to national technical means (NTM) to carry out independent aerial and other observation.

What makes things very difficult in the case of technology transfer control is that a high level of national cooperation cannot necessarily be counted on. It is likely that a country under suspicion may not have the social and industrial infrastructure normal in industrially advanced countries. One may think of Iraq under Saddam Hussein's regime and wonder if, under normal conditions, somebody within Iraq who is concerned about industrial safety could have gone to the central registry and asked for detailed information about uranium-handling facilities in Al Tarmiyah or Al Sharqat (both of which turned out to be secret uranium enrichment plants). A similar fear about the People's Democratic Republic of Korea may turn out to be justified, causing one to wonder if nuclear sites (among them the plutonium separation facility in Yongbyon) could produce detailed material accountancy and personal radiation records. The world already knows how things have been 'un'-managed in the former Soviet Union's weapons and submarine facilities. Things may very possibly be worse in some yet unnamed chemical plant in Libya.

One may try to draw some sort of parallel with official development aid (ODA), and wonder why in so many cases basically well-intentioned government aid seems to be wasted either on misdirected or useless works, or very often through alleged corruption, without ever reaching the people in real need. The simple explanation often given is that countries that can make maximum use of outside aid are those equipped with a bureaucracy and an infrastructure, and that such countries do not need aid. One may recall the aid that was provided (primarily by the United States) to Europe and Japan after the end of the Second World War. These were the countries that could make the best use of aid for reconstruction and development, and that undertook to repay the aid. Those with direct experience in trying to make current arrangements for ODA realize that the regions involved are often without a normal societal infrastructure. In a country where industrial standards are virtually non-existent, small nationally produced household items are often incompatible: for electric lightbulbs to fit into sockets there is a requirement for manufacturers to observe the same industrial standards.

This aspect of confusion adds to the difficulties already mentioned in carrying out effective export control. It is somewhat similar to the argument about global environmental control. If, for example, carbon dioxide contributes to the greenhouse effect, and any one of the many mathematical models is correct in predicting the dire consequences of carbon dioxide accumulation, it is prudent in the name of civilization to require developing countries of the world to use less untreated fuel. But there can be double trouble in this recommendation. An obvious counterargument is that industrially advanced countries that have hitherto been contaminating the global environment have no right to ask their developing brothers to refrain from fuel combustion. In addition to this general argument, one may add that cheap, easy-to-use, less sulphurous and cleaner energy such as natural gas and light oil should be reserved for use by the developing countries, while the industrially advanced should do more to explore high-tech energy such as nuclear fission and/or nuclear fusion to prepare the way for large-scale energy consumption in the twenty-first century.

Conclusion: what should be done

The above discussions all seem to converge on one point: the future of science and technology, which may be an appropriate subject for joint consideration in the post-Cold War world. There is a need to determine the direction of science and technology for the coming decades. Just as people in the postwar United States evaluated science, the endless frontier,[8] the post-Cold War world needs a joint effort by scientists, engineers, and political and social scientists, as well as those in political leadership, to think about the coming role of science and to provide a blueprint for international joint research and development, international control against abuse, and international encouragement for those willing and able to undertake scientific research as their lifetime career. It is too early to give even a rough description of how these international arrangements would work. The list of problems included in this and other writings about post-Cold War restraints on proliferation and handling of nuclear and chemical weapons will be a mere litany of complaints unless ways are found to derive positive results from awareness of the difficulties.

Figure 1: Leading arms importers of the world (1989)

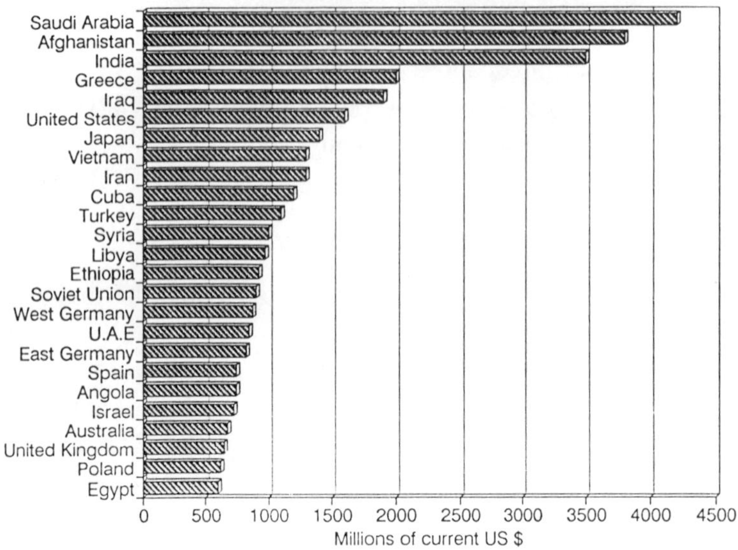

Source: *World Military and Arms Transfer, 1990*, US Arms Control and Disarmament Agency

Figure 2: Japan's technology imports (1950–1990)

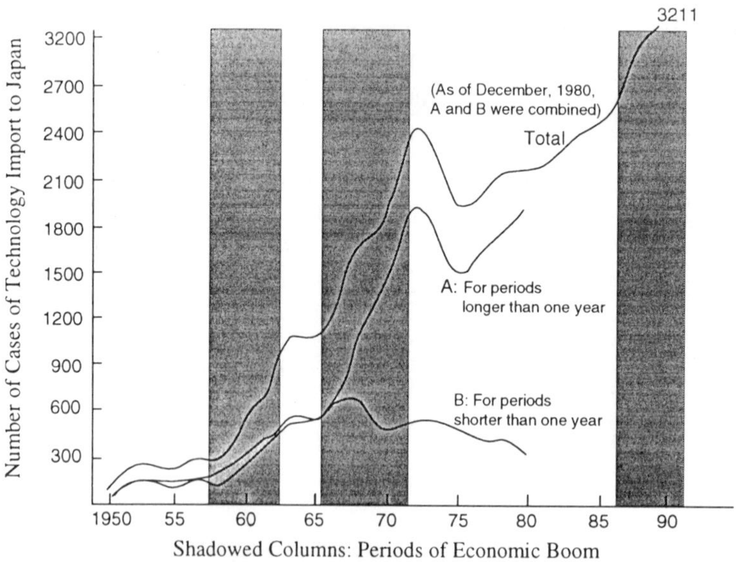

Source: Japanese Government Statistics

Table 1.1: Major US defence contractors (over a 5-year period)

Firm	Industry	Prime contract $ million	RDT&E $ million	% of RDT&E/ R&D	% of prime contract/ total sales
McDonnell Douglas	aero	8,030	1,413	70	53
General Dynamics	aero	7,226	484	54	70
General Electric	gen	6,449	1,073	48	13
Lockheed	aero	5,224	1,578	75	43
Rockwell	aero	4,683	791	65	32
General Motors	auto	4,579	559	11	4
Boeing	aero	4,178	1,219	62	23
Raytheon	comm	3,717	542	67	47
United Technologies	aero	3,665	379	29	20
Martin Marietta	aero	3,163	1,417	88	61
Grumman	aero	2,965	706	98	80
Litton Industries	comp	2,112	144	64	48
Westinghouse	gen	1,956	469	71	16
Tennoco	oil	1,952	2	n/a	19
UNISYS	comp	1,932	362	34	18
Westmark	comm	1,904	249	60	25
Honeywell	instr	1,753	306	49	23
IBM	comp	1,554	441	9	3
LTV	misc	1,439	237	91	18
Texas Instruments	comp	1,272	147	23	20
TRW	comp	1,135	621	73	17
ITT	comm	1,080	169	25	5
Allied Signal	aero	994	25	6	8
FMC	chem	923	78	35	28

Source: This table, and Tables 1.4 and 1.5, reprinted by permission of Harvard Business School Press from *Beyond Spinoff, Military and Commercial Technologies in a Changing World*, by John A. Alic, Lewis M. Branscomb, Harvey Brooks, Ashton B. Carter and Gerald L. Epstein. Boston: 1992, pp. 89 and 192 (here p. 192). Copyright © 1992 by the President and Fellows of Harvard College.

Key: RDT&E: Research, Development, Test and Evaluation; comp: computing; comm: communications.

Table 1.2: Major arms producers of Western Europe (1989, in US$ millions)

Company	Country	Arms sales	% of total sales	Profit total
British Aerospace	UK	6,300	42	3546
Thomson S.A.	France	4,320	36	78
Daimler Benz	Germany	4,260	10	3622
Thomson-CSF	France	4,120	78	412
DASA	Germany	3,930	53	-58
Dir. de Const. Navales	France	3,630	100	—
GEC	UK	2,800	20	1118
IRI	Italy	2,230	5	1178
Dassault-Breguet	France	2,200	72	46
Aerospatiale	France	2,190	44	32
MBB	Germany	1,840	55	16
CEA Industrie	France	1,820	35	345
Lucas Industries	UK	1,640	45	313
EFIM	Italy	1,510	40	-18
INI	Spain	1,290	8	693
FIAT	Italy	1,280	3	2410
SNECMA Groupe	France	1,260	37	46
Rolls Royce	UK	1,220	25	382
Oerlikon-Buhrle	Switzerland	1,040	36	-25
GIAT	France	1,020	97	-107
Aeritalia (IRI)	Italy	1,020	55	41
Nobel Industries	Sweden	950	27	214
Siemens	Germany	900	3	839
Matra Groupe	France	870	25	95

Source: Michael Brzoska and Peter Lock (eds), SIPRI, *Restructuring of Arms Production in Western Europe* (Oxford: Oxford University Press, 1992), Appendix A.

Table 1.3: Major defence procurement in Japan

Firms	Defence sale (¥ billion)	% of defence in total sale	Employed in defence
Mitsubishi Heavy Industry	431.9	17	7,000
Kawasaki	185.1	19	5,000
IHI	116.4	14	3,000
Mitsubishi Electric	150	6	2,000
NEC	100	3	n.a.
Toshiba	60	2	2,000
Fuji Jukou	43	5	2,800
Nihon Seikou	35	23	730
Komatsu	23	4	400
Hitachi	50	1	1,000
Daikin	20	5	250
Fujitsu	37	1.5	1,100
Nissan	22	0.5	900
Mitsubishi Precision	18	70	1,200
Nihon Koki	17	85	700
Hitachi Ship Building	18	6	800

Source: *Toyo Keizai,* 13 February 1993, special issue on Japan's defence planning.

Table 1.4: Status of defence industries in Japan (1990) (in millions of yen)

	Defence production (A)	Total production (B)	Ratio A/B
Ships	161,889	2,306,938	7.02
Aeroplanes	551,578	751,978	73.35
Vehicles	25,209	44,061,569	0.06
Munitions	450,602	451,569	99.89
Communications	317,179	54,739,597	0.58
Oil products	63,049	7,974,635	0.79
Fibres, clothes	10,157	12,398,722	0.08
Medical products	6,239	5,172,984	0.12
Foodstuffs	39,781	33,106,817	0.12
Other	133,735	163,872,955	0.08
TOTAL	1,759,418	324,959,179	0.54

Source: John A. Alic, Lewis M. Branscomb, Harvey Brooks, Ashton B. Carter and Gerald L. Epstein, *Beyond Spinoff, Military and Commercial Technologies in a Changing World* (Boston, MA: Harvard Business School Press, 1992), p. 192.

Table 1.5: A comparison of national R&D spending in 1988 (in billions of US dollars according to OECD purchasing power parities)

	R&D spending	Defence R&D	R&D/GDP %	Non-R&D/GDP %
United States	133.7	43.0	2.7	1.9
Japan	47.0	0.4	2.6	2.6
FR Germany	24.6	1.1	2.8	2.7
France	17.5	3.9	2.3	1.8
United Kingdom	17.0	3.3	2.2	1.8
Italy	9.1	0.6	1.2	1.1
Canada	6.4	0.3	1.4	1.3
Netherlands	4.3	0.1	2.3	2.1
Sweden	3.6	0.4	2.9	2.6
Switzerland	3.2	0.1	2.9	2.9
non-US total	132.9	10.2	2.3	2.1

Source: John A. Alic, Lewis M. Branscomb, Harvey Brooks, Ashton B. Carter and Gerald L. Epstein, *Beyond Spinoff, Military and Commercial Technologies in a Changing World* (Boston, MA: Harvard Business School Press, 1992), p. 89.

Table 1.6: International control regime over technology transfer

	Treaties and/or agreements	International organization	Verification methods	International understanding
Nuclear	Non Proliferation Treaty (NPT) 1970	International Atomic Energy Agency	Nuclear material control, safeguards	London Guidelines (1978)
Chemical	Geneva Convention (1925); CW treaty (1993)	Organization under way in The Hague, Netherlands	Challenge inspection	Australia Group 1985
Biological & toxin	Treaty (1975)			Australia Group
Ballistic missile				MTCR (1987)
Military in general				COCOM (1949)

Notes

1 US President Jimmy Carter announced on 23 January 1980 that 'Any attempt by an outside force to gain control of the Persian Gulf region will be regarded as an assault on the vital interests of the United States of America, and such an assault will be repelled by any means necessary, including military force'. This statement later became known as the *Carter Doctrine*.

2 Thomas B. Cochran, William M. Arkin and Milton M. Hoenig, 'US Nuclear Forces Capabilities', *Nuclear Weapons Databook* (Cambridge, MA: Ballinger for Natural Resources Defense Council, 1984).

3 The US Arms Control and Disarmament Agency publishes *World Military Expenditures and Arms Transfers* on an annual basis. Other annual publications include the International Institute for Strategic Studies' *Military Balance* and Stockholm International Peace Research Institute's (SIPRI) *World Armaments and Disarmament*.

4 Tom A. Zamora, in 'New Jobs for Old Labs?' (*The Bulletin of the Atomic Scientists*, November 1992), gives some detailed descriptions of the problem in the US weapons laboratories.

5 J. A. Alic, L. M. Branscomb, H. Brooks, A. B. Carter, and G. L. Epstein in *Military and Commercial Technologies in a Changing World* (Cambridge, MA: Harvard Business School Press, 1992) have an extensive discussion of the differences in basic structures between defence and

civilian industries. Ken Aedelman and N. R. Augustine, in 'Defense Conversion: Bulldozing the Management' (*Foreign Affairs*, Spring 1992), and D. Berenstein and K. Smith in 'Collaborative Project on Soviet Defense Conversion' (Stanford Centre for International Security and Arms Control) describe Russian cases.

6 An interesting discussion of this incident from the point of view of US trade policy may be found in Raymond Vernon, Debora L. Spar and Glenn Tobin, *Iron Triangles and Revolving Doors; Cases in US Foreign Economic Policymaking* (New York, London: Praeger, 1991).

7 Because the case was switched to a damages suit and since the government won the case in that damages were not awarded, it lost an opportunity to appeal in terms of the interpretation of the law.

8 In 1944 Vannevar Bush wrote, at the request of President Franklin Roosevelt, a 192-page report outlining the role of science in the postwar United States. The report discussed the postwar reorganization of science in the government and provided a blueprint (which was given to President Truman). *Science, the Endless Frontier* is often considered a far-sighted vision of (organized, or big) science as it emerged during the war, primarily through instruments such as the Manhattan Project.

CHAPTER 2

THE DEMAND FOR NUCLEAR TECHNOLOGY AND ARMS: FUTURE POLITICAL IMPLICATIONS

YOSHIAKI NAKAGAWA

Introduction

Although, the Cold War is over, Iraq's military action, the war in former Yugoslavia and the continuing turmoil in Somalia, Angola and Cambodia demonstrate that in the future the world will not be without conflict. Racial, religious and ethnic differences will continue to be sources of strife. Political revolutions are bound to erupt as societies evolve in the new global structure. Over the past 45 years, wars and conflicts around the world have resulted in 40 million deaths. There is no reason to anticipate that historical disputes over political boundaries will cease. In addition, the economic gap between nations seems likely to widen, causing further instability.

One of the most critical issues for the stability of the world is nuclear proliferation. Despite the progress made in nuclear disarmament talks between superpowers at the end of the Cold War the nuclear threat is still great. One particular area of concern is the Third World, which has the greatest number of potential nuclear powers. Currently, Third World military expenditures total nearly $200 billion per year, which is approximately 5 per cent of their gross domestic product (GDP), a proportion comparable to that of the United States. Such spending is alarming, not only in view of world security, but also for its effect on the economic development of these countries.

It is a common claim that although the Cold War had the effect of containing conflicts between regional rivals in the Third World, it turned these regions into an ideological battleground. Thus, these countries have inherited factors of instability. The US–Soviet rivalry was undeniably a contributing factor, but the underlying causes of conflict in the Third World existed even before the Cold War began, and they will almost certainly continue even though it has ended. If certain Third World countries gain nuclear weapons capability, it will have adverse effects on world security as a whole.

In addition to the continuing disorder in the Third World, potential strife may erupt in the states of the former Soviet Union. Thus the world of the future will not be so different from the world of the past; conflicts within and between nations will not disappear. Also, the increasing threat of nuclear proliferation will contribute to a very grave future situation.

Over the past three decades, efforts have been made to limit or to abolish non-conventional weapons and weapons of mass destruction such as nuclear, biological and chemical weapons. Ballistic missiles capable of delivering weapons of mass destruction have also been a source of concern, as they are indispensable for completing an effective non-conventional weapons system. The 1968 Nuclear Non-Proliferation Treaty (NPT) and the 1972 Convention on Biological and Toxic Weapons have unfortunately done little to slow the spread of weapons in these categories. Consequently, today 23 countries possess or have the capability of producing at least two of these four categories of weapons (nuclear, biological, chemical, and ballistic missiles). A nuclear potential is the single most common; indeed, most of them either already possess or are capable of producing nuclear weapons. These countries can be categorized in the following manner:

First-tier countries. These are the 'mature'[1] five UN Security Council permanent member countries and the G24 countries, all of which are relatively stable. These are mainly the modern 'like-minded' industrialized countries and advanced democracies which support non-proliferation, at least officially.

Second-tier countries. These are relatively stable countries whose threat to international security is not of immediate critical concern. This group comprises Ukraine, Belarus, Kazakhstan, Israel, South Africa, Argentina, Brazil and Taiwan.

Third-tier countries. These are relatively stable countries which could become 'premature'[2] or 'first-strike'[3] nuclear countries. They are of particular concern to the first-tier countries. This group comprises North Korea, Iraq, Syria, India, Pakistan, Egypt, Iran, Libya, South Korea and Cuba.

In addition, it is possible that there may be other countries we do not yet know of which already possess or are developing such capabilities.

This chapter focuses on why a potential nuclear state may feel the need to possess nuclear technology and its own nuclear arms. Further, it will explore the global impact of nuclear proliferation.

Incentives for nuclearization

The world is in the process of moving from the old energy-producing age of coal and oil into the nuclear era. This powerful new energy source is attractive to many countries, particularly developing nations. A prerequisite for the effective use of nuclear energy is a high level of technology and its possession is regarded as the mark of a modern industrialized state. To attain this high technological level, however, huge investments of national resources are necessary, and this usually requires extensive government support.

Though nuclear power is arguably a more efficient and economical method of producing energy, its political allure is declining in some areas as a result of the anticipated rising costs associated with safety and environmental protection. In many modern industrialized countries, the use of nuclear power is challenged by domestic groups, raising political and social questions for the government. Yet for many governments in the Third World, the development of nuclear energy is seen as a means of heightening their country's international prestige.

Overall, it is difficult to deny nuclear expertise and technology to any country because that would mean preventing the development of a part of industry which could enable the country to become competitive in the global market.

Yet nuclear technology is by nature a typical 'dual-use technology' and there is no clear technical discontinuity between a peaceful nuclear programme and a programme of nuclear weapons development. The problem for the developed states lies in how to promote the peaceful use of nuclear energy while halting both vertical and horizontal nuclear weapons proliferation.

The most powerful factors behind a nuclear weapons programme may be national security concerns, reflecting a country's perception of external threats, but political and prestige considerations can also be important. In considering the issue objectively, it may be said that possession of nuclear arms could provide a country with the following strategic capabilities:

(1) *To counter a massive conventional attack*: this capability is extremely important for a country whose population and conventional forces are much smaller than those of its adversary. For example, this may be a particularly telling factor for Pakistan, Israel and Ukraine;
(2) *To offset any political or geostrategic superiority* that other powers with large territories and populations may enjoy. Nuclear arms may have been an important factor in equalizing the UK and France, on the one hand, and the Soviet Union on the other. It signalled a British and French belief that there were four great powers, not two superpowers (the US and the Soviet Union) plus others;
(3) *To deter a nuclear attack*, provided that the nuclear force can withstand a first strike from an adversary. Some states, however, may not have invulnerable nuclear forces.

At present, the incentives to obtain nuclear weapons also reflect the international political value of nuclear weapons and the prestige they may be thought to bring.[4] Basically, a country might wish to engage in a peaceful nuclear programme to show it can achieve a significant level of technological advancement. However, it is not a big step to move from a peaceful nuclear programme to a programme of weapons development.

On the other hand, a country that has obtained or is considering obtaining nuclear weapons may face strategic disadvantages, such as:

(1) the need to allocate a significant part of its national resources for nuclear weapons programmes, including well-trained scientists and engineers, research facilities and capital, for example. This can be at the expense of economic growth or the welfare of the nation. To plan and maintain a military posture that can handle a hypothetical nuclear exchange and maintain an effective deterrence capability is particularly expensive. As Russia is discovering, even dismantling nuclear weapons is expensive. This disadvantage may apply to all nuclear-armed countries, be they 'mature' nuclear states (such as Russia) or those defined by the NPT as 'premature' nuclear-armed countries;
(2) the possibility that the North or countries in the Nuclear Suppliers Group (the London Group) may impose sanctions or take other hostile actions to prevent a new nuclear power from forming. In

order to avoid the threat of sanctions, some countries adopt the strategy of keeping their nuclear weapons programme at a point of only potential capability. They thus avoid declaring nuclear weapons possession, and escape international detection by refraining from explosive tests and/or by accepting inspections by the IAEA.

A comparison of Japan or Germany with India shows a clear difference in strategic and political calculations with respect to non-proliferation. Japan and Germany have realized that from both a military and a security standpoint, the possession of a nuclear weapons system is beyond their political capability, owing to the political, economic and security costs that they would have to bear. Though both countries already possess the technological potential, they recognize that acquisition of nuclear weapons would be politically expensive and would contribute very little to their own security.

India is an example of the opposite case, a 'premature' nuclear-armed country. India detonated a nuclear device in 1974. The Indian government apparently decided that nuclear weapons development and the expected resulting increase in international prestige would be worth the economic and political costs incurred. However, it has not proceeded to formal deployment of a nuclear force, a fact that reflects the political and technological pressures which it feels.

The impact of nuclear proliferation

We can anticipate two major political consequences from nuclear proliferation: a significant change in regional stability in areas where potential or existing conflict exists, and a significant change in the non-proliferation strategies of the members of the Nuclear Suppliers Group.

The impact of nuclear proliferation on regional stability

Technologically, what is the possibility that a nation which desires to produce nuclear weapons may be able to do so? It is possible to explode devices which are made with reactor-grade plutonium, though they may have an uncertain yield. However, such devices would be unreliable and would deteriorate over time. There are also problems in cooling the reactor-grade plutonium 'pit' of the weapon and in making a radiation shield to protect the operative. Practically speaking, reactor-grade plutonium devices have poor prospects for deployment. Yet for a country which is just beginning to acquire nuclear weapons technology, this kind of crude device might be the first bomb to make for economic and technical reasons.

The impact on regional security of a country possessing such a device would be psychological rather than military. The dimension and weight of the crude plutonium device would make it too large for delivery by ordinary military transportation vehicles such as fighter-bombers or ground-to-ground missiles. It would require a cargo carrier (such as a C-130 or Boeing 737) or a heavy cargo vehicle. Considering the factors of unreliability and handling difficulties, the device could not be regarded as a viable weapon. Yet a successful test by detonation could allow the country to claim that it had significantly challenged the existing world order, and had attained the same level as the other (nuclear) world powers. The consequences of such an achievement by a Third World country should not be ignored. It is, however, doubtful that such a device would have significant military meaning, or would pose a real threat to other countries.

In order to have military significance, a nuclear device must be of a quality to meet the minimum standards for a weapon. Such standards include the reliability and safety of detonation control, consistency of yield, transportability, ease of operation, and manufacturability. Once a weapon can be produced that meets the minimum standards in these areas, a country can progress its nuclear weapons programme to the next step of making an efficient, small, advanced weapon. Then in order to make its nuclear capability operational, the country would need modern fighter-bombers and/or surface-to-surface missiles, and an appropriate command and control machine. The entire system must be created and kept highly operational in order for it to have military significance and a deterrent capability.

The total effort needed to achieve this makes a nuclear weapons programme a very expensive undertaking. Yet many states could obtain some degree of nuclear warfare capability, probably in the form of a system with a very limited number of warheads and delivery capability/vehicles, should they be willing to pay the economic and political costs. There are variations on the approach outlined above such as Iraq's programme, which sought to use the more expensive, highly enriched uranium (HEU) as the fissile material instead of plutonium.

A region in which a 'premature' nuclear-armed country emerged would of course be a more volatile strategic environment than if no such capability existed.[5] The strategic stability of such a region would essentially depend on the vulnerability of warheads and delivery vehicles, the vulnerability of command and control systems, a nuclear state's own threat perceptions and speculation on the likelihood of war, and its strategic rationale for possessing nuclear arms.

Nuclear arms owned by a 'premature' country are likely to be more vulnerable than those of 'mature' powers. Even with a favourable geographic environment, it is very difficult to prevent detection of even a limited number of warheads and delivery vehicles over the long term. Delivery vehicles in particular are hard to disperse effectively. Tactical fighter-bombers which have been modified for nuclear missions must remain stationed near air strips, and missiles are most likely to be kept in the fixed hardened bases at the early stages of their deployment owing to the cost and technical difficulty of mobile capability. As a result, nuclear arms may be deployed in a few well-known, and thus vulnerable, locations. If this is so, it is likely that a 'premature' nuclear country will feel a strong incentive to use its nuclear arsenal early in a conflict, before the enemy can destroy it.

The same pre-emptive logic is applicable to the country's command and control system. Depending upon the vulnerability of the overall system, a country might have an incentive to use its nuclear weapons before they are made unusable by the enemy. Particularly in 'premature' nuclear-armed countries, where the leaders may be concerned about their own survival and where the key nodes of communication in the national system may break down (as they are not fully protected or constructed with enough redundancy), this is a dangerous situation.

Threat perceptions and calculations of the likelihood of war can also affect a proliferator state. In 1967 Israel attacked its Arab neighbours in a classic conventional pre-emptive strike. Even if Israel had possessed a significant nuclear force at the time, it might still have felt the need to attack, since it could have been deterred from relying on its nuclear systems by what it regarded as the Arabs' implicit nuclear guarantee from the then Soviet Union. Arguably, Israel sought to build ever larger ranges into its missiles in the 1980s so as to deter the Soviet Union if necessary as well as the Arabs. Extended nuclear guarantees, even of an implicit nature, will stimulate reactions on the part of targeted states.

Any strategic rationale for a nuclear system will have an effect on regional stability. A rationale which points to the early and initial use of nuclear weapons may either provoke alarm or suffer from credibility problems. Israel has not made public any strategic rationale for its nuclear forces, but it is significant that such forces did not deter the Egyptian–Syrian attack of 1973.

Political and military stability in many regions would be severely impeded by nuclear proliferation, and overall it would be practically impossible for a 'premature' nuclear-armed country to ensure its own peace and security, because of the regional instability that would be likely to ensue.

The impact on nuclear non-proliferation strategies

The traditional approaches taken to block a country's nuclear development (usually through the control or denial of nuclear technology) are becoming less valid. The situation on the Indian subcontinent, South Africa's recent revelation that at one point it possessed nuclear weapons but did not actually perform detonation tests, and North Korea's suspected nuclear programme show the limits of the current nuclear non-proliferation regime. We must recognize that the success of nuclear non-proliferation must rely on continuous political persuasion of emerging countries by means of both 'sticks' and 'carrots'.

Historically speaking, after India's experimental nuclear explosion in 1974 the international non-proliferation regime was restructured to make safeguards and export controls the major political and security tools of the 'mature' nuclear-armed states. The application of a technology-denial strategy or a technology-transfer control strategy was the key element of the non-proliferation system, and the list of prohibited export items was gradually lengthened and expanded over time. These measures, however, achieved very little in preventing proliferation, though it is anticipated that they should have some restricting effects on nuclear weapons programmes in countries such as India and Pakistan. However, it is more likely that these measures will affect the peaceful nuclear programmes of these countries and may encourage efforts on the part of the countries concerned to pursue the independent/self-sustained development of nuclear technology which would eventually be useful in a nuclear weapons programme. Twenty years have passed since India detonated a nuclear device, and the number of countries suspected of pursuing nuclear weapons programmes is increasing steadily. Technical controls can buy time but they cannot resolve the proliferation problem, nor can they stem independent/self-sustained technology developments in many countries.

By their very nature, the IAEA safeguards which were designed to prevent or to stop nuclear proliferation could not prevent the proliferation of nuclear weapons potential. The safeguards could only prevent the conversion of this potential into actual weapons production. As pointed out above, a nuclear explosive device which uses reactor-grade plutonium (basically, the by-product of commercial electric power generation) can be detonated.

We must realize that preventing the proliferation of important dual-use technologies such as nuclear technology is a political and, at the same time, an economic problem rather than a technical one.[6] Technical means, such as safeguards, are not effective in solving what are basically political and economic problems. The most that technical means can do is to assist political means when they alone cannot achieve the desired results.

Strategies to promote non-proliferation

Lessening the nuclear incentive

Nuclear weapons and nuclear weapons potential carry international political weight, as they may provide leverage for a country in international disputes or conflicts. In peacetime as well, they can enhance prestige. Therefore, in order to promote demand-side disincentives, we must begin by lowering the political value of nuclear weapons. The fundamental approach to non-proliferation should begin with lessening, even denying, the legitimacy of nuclear weapons possession in the international political arena.

During the Cold War, this was an unrealistic ideal. But now that the Cold War is over, it may have possibilities. The first-tier countries, especially the 'mature' nuclear-armed countries, are struggling to redefine the meaning and role of nuclear weapons in their strategic planning. The meaning of deterrence has changed: it is no longer a clear-cut matter of superpower protection of itself and its allies against another superpower.

The START I and II agreements clearly show that present global political conditions allow the US and Russia to get rid of many of their nuclear weapons. Similar political conditions exist for the UK, France and China, so it follows that they must also rethink the meaning and role of their nuclear weapons. As START I and II are implemented, these three countries will also dismantle at least a certain number of their nuclear weapons. This is an important second step.

In order to advance the goals of non-proliferation, it would be a mistake for any state to emphasize the political utility of nuclear weapons. The consolidated views of the first-tier countries regarding the legitimacy of possession, let alone use, of nuclear weapons will define the future political value of these weapons.

Promoting demand-side disincentives

Theoretically, disincentives for nuclear proliferation on the demand side derive from strategic considerations: a country must analyse the costs and benefits of nuclearization. If a potential nuclear state realizes that the political and security benefits it might gain will not compensate for the political and economic costs, then the goal of non-proliferation can be achieved.

Clearly, political and security benefits are relatively few or even non-existent in contrast to the huge costs of becoming a nuclear-armed country. Japan and Germany can serve as examples of the success of non-proliferation efforts and of a strategically rational consideration of the nuclear weapons option, though admittedly their example may not be appropriate for all countries.

Countries opposed to proliferation must also take steps to ensure that more states reach the same conclusion as Germany and Japan: that possession of nuclear weapons cannot enhance their security. Positive and negative security guarantees on the part of nuclear states have a role to play, as does the promotion of cooperative security arrangements as an alternative to confrontation and deterrence.

A particularly firm stand must be taken against countries like North Korea, which are tempted to use the threat of acquiring nuclear weapons as a source of wider political leverage. There is also concern that such countries, in the event of their acquiring a nuclear weapons capability, could well quickly resort to making nuclear threats. To an extent, the international community may be able to 'buy off' states like North Korea with economic help provided in exchange for non-proliferation, but there are clearly limits to the resources which can be made available. Other tools may need to be mobilized.

Political tools for promoting disincentives

Non-proliferation is one of the most important international political goals. Yet it is also an endless process. If any potential nuclear country believes that it can challenge the modern industrialized countries or their 'world order' by acquiring nuclear weapons without penalty, proliferation may be inevitable. To promote non-proliferation and maintain world peace and security, clear objectives must be defined, and we must make use of both big 'carrots' and formidable 'sticks'.

The objectives should be as follows:

(1) To make the potential nuclear-armed country realize the political costs of losing friends in the world, the direct economic costs of carrying out a nuclear weapons programme, the indirect economic costs of possible sanctions and lost benefits, and the military risks of nuclear weapons possession;
(2) to encourage the country to seek alternative means of achieving enhanced national prestige, such as sustained economic growth;
(3) to prevent the acquisition of militarily significant nuclear weapons. If prevention efforts fail, all other means must be used to encourage voluntary abandonment of a nuclear weapons programme, particularly in cases of 'countries of political concern'. In this regard, the recent revelation of South Africa's voluntary dismantling of its nuclear weapons provides an example. Should voluntary measures fail, more drastic means of persuasion may have to be considered by the first-tier countries.

The political 'carrots' to be used are:

(1) economic/military aid;
(2) security guarantees;
(3) arms control measures.

The political 'sticks' are:

(1) export controls;
(2) deterrence;
(3) economic sanctions;
(4) preventive warfare (to destroy the political will to produce/use nuclear weapons, as well as the actual production capability).

Other political tools, namely defence, may be useful for self-defence against a definite threat, but they are not effective in preventive non-proliferation.

It is clear that there are specific considerations relevant to each of the political tools above. Adequate security guarantees would be hard to agree for countries such as Iran and Iraq whose poor human rights records make them liable to sanctions from the international community.

Also, as discussed above, extended deterrence can have an unsettling effect on the regional strategic situation. Yet, should the 'mature' nuclear country providing extended deterrence think of making a retaliatory strike against an aggressive nuclear country, it is likely that strong domestic objections against either nuclear or conventional retaliation would arise: civilian casualties would probably occur, and from the beginning the people might not support their country's involvement in this kind of foreign military action. Though this opposition might not have an immediate effect on the region in question, it would eventually serve to undermine the credibility of the 'mature' country as a force providing extended deterrence in the region.

In a region where a state is aggressively seeking change, arms control measures are not very attractive as a substitute for a military build-up and defence modernization. Arms control measures themselves do not normally include provision for automatic punishment in the event of their violation. Without the use of other political or military means to back them up, arms control measures amount to little more than words on paper.

Export controls are infamous for the difficulty of their enforcement. It is comparatively easy for the large number of suppliers to bypass them, and the overlap of military and civilian technologies and projects makes distinctions difficult. As stated earlier, export controls can also be an incentive for a country to pursue independent technology development.

Deterrence requires a rationality which puts a significant value on human welfare and life. It is thus not absolutely reliable. In any instance, it may not be a wise policy to trust the rationality of a 'country of concern' which may be attempting to acquire nuclear weapons and ignoring the adverse political and economic implications.

Economic sanctions may be an effective method of discouraging proliferation, but a rather long time-span is needed to effect change in the political course of action of an emerging nuclear-armed country. Sadly, what sanctions affect first is the living standard of the civilian population. Sanctions are usually accompanied by a high death rate among infants and the elderly. To avoid creating such a tragedy, non-proliferation efforts should instead centre on instant and effective measures. Therefore, sometimes the option of waging preventive war may be preferred.

A decisive preventive war can be highly effective. Yet considering the costs and risks involved, this option is not likely to be used extensively in the future. It is more likely that 'surgical' attacks with the limited objective of destroying actual nuclear weapons and production facilities will be employed because of the lesser risks and costs involved.

It is apparent that there are shortcomings involved in any of the above political means. Nonetheless, it is imperative that we set objectives, consider solutions to potential problems, and employ the political means available as consistently and effectively as possible.

For the like-minded first-tier countries (hopefully all of them), a proposal for a grand strategy of nuclear non-proliferation could consist of one carrot and one stick backed by a firm commitment to enforcing non-proliferation, together with self-restraint in using or possessing nuclear weapons. To preserve a minimum level of security and safety, at least one politically meaningful and achievable intermediate objective would be to prevent countries from obtaining militarily significant nuclear warheads attached to ballistic missiles. This type of nuclear force could have a lethal effect on the civilian population of the industrialized countries. In order to prevent the proliferation of both nuclear weapons and ballistic missiles, the first-tier countries should, first, reinforce the existing 'Missile Technology Control Regime'(MTCR). This may mean expanding the list of members to include all supplier countries, yet this must be done with discretion, for MTCR members share technically sensitive information which could be used by certain countries to raise their present technology level. Therefore reinforcement of non-proliferation would require strict control, particularly over range-lengthening technology and its near relation, payload-increasing technology.

The Third World suppliers may bypass the regime's non-proliferation efforts despite their inclusion in the MTCR. Should this be the case, a second group of policy measures would be necessary, involving an international agreement on a Comprehensive Test Ban (CTB). A CTB would be useful in preventing or significantly delaying the development of small, lightweight and militarily significant warheads which could be carried by missiles. There are significant technical gaps between making missile nuclear warheads which can be effectively deployed and the first generation of crude nuclear-explosive devices. In order to bridge the gap, extensive research and development efforts, including numerous test explosions, are necessary. Without actual detonation, even state-of-the-art technology cannot ensure the secure and safe performance of nuclear warheads in the field.

Should these two major non-proliferation methods fail to prevent a country from acquiring nuclear warheads and ballistic missiles, the next step would be 'non-proliferation enforcement'. This would entail invoking economic sanctions or even launching a preventive war with close policy coordination and burden-sharing measures linking the international community.

Conclusion: prospects for success

The prospects for non-proliferation can be seen as fairly good if enactment of the MTCR and employment of the CTB can be carried out effectively. The major reason for optimism is that non-proliferation problems are political in nature, and therefore the emergence of a new nuclear-armed country is not the end of the world. New political regimes could be devised, if with difficulty.

The most urgent item on the non-proliferation agenda at present is North Korea. On 29 May 1993, North Korea test-launched its new 1,000 km-range ballistic missile while in the midst of NPT negotiations with the United States and other countries. Should North Korea's emerging nuclear weapons capability be seen by other nations as having successfully challenged the established world order and the superpower stance of the United States there would be no way to prevent a second, or third, or fourth North Korea-type country from challenging non-proliferation. Therefore, the way in which the situation in North Korea is handled by the international community will have a significant effect on the success of current non-proliferation efforts.

We should respond to this challenge not with words alone, but also with action, with 'big carrots' and 'strong sticks'. This is not just to answer North Korea's challenge, but to set an example to all other potential nuclear-armed countries. Only by doing so can we hope to stem nuclear proliferation.

Notes

1 Such countries can be considered 'mature' in that they possess sufficient knowledge, experience and technical background to operate their own nuclear weapons.

2 By 'premature' I mean that such countries have no or only negligible knowledge, experience and technical background to operate their own nuclear weapons. In short, considering their social conditions and development, they are taking on nuclear armament too quickly.

3 'First-strike' means the country is without second-strike capability. This is due to the inherent vulnerability of its nuclear forces (including nuclear weapons and the associated command and control systems).

4 Historically, I view the nuclear armament programmes of the UK and France after the establishment of NATO (1949) as connected to their relations with the US, and prestige considerations, as well as security concerns. Nuclear weapons have only twice been used in conflict since their development and their international political value is not necessarily directly related to national security concerns.

5 Some argue that nuclear weapons have been a stabilizing factor in regional conflicts: For example in the Arab–Israeli conflict it can be said that the Arab side has been restrained by its knowledge of Israel's nuclear capability. However, in my view, Israeli nuclear capability may have actually ignited the fuse of nuclear proliferation in the Middle East. I place five of Israel's neighbours in the third-tier country category. It is obvious that the Arabs will refuse to be subject to Israel's nuclear dominance indefinitely.

6 There may be other aspects, such as the 'moral' position. Third World countries may claim that the non-proliferation policies of the North are designed simply to maintain the status quo which favours the North. They may say that the NPT and other such international agreements are to help those who already possess nuclear weapons to maintain their superiority. In this chapter, I have concentrated on political and military aspects of non-proliferation. However, it is clear that with the end of the Cold War and the threat of nuclear proliferation, it is not only world security that is in question, but also the very existence of human beings, both in the North and in the South.

CHAPTER 3

THE THIRD WORLD DEMAND FOR CHEMICAL AND BIOLOGICAL WEAPONRY AND BALLISTIC MISSILES

MARTIN S. NAVIAS

Continuing demand

The demand for ballistic missiles and biological and chemical weapons among developing states continues to be strong. Indeed, in a number of important instances efforts to deploy such systems may be intensifying. Despite the reduction of superpower competition, the significant level of attention being paid to arms control by the industrialized world, and the declarations by many developing countries that they are committed to curbing proliferation of ballistic missiles and chemical and biological weapons, there is no indication that the proliferation problem is fully under control. The implications of this for regional and international security are unfortunate, especially with regard to the highly volatile Middle East and South Asian regions and potentially problematic North Asia. Patterns of global proliferation, rationales for acquisition and the technical attractions of these systems, all of which have remained relatively constant since before the end of the Cold War, suggest that the demand for such systems will continue in some regions into the foreseeable future and that global arms control will not be easy.

There are several Third World states believed to have deployed or to be presently developing chemical weapons. Probable chemical weapons states (that is, those named by US government sources on the record as deploying, possessing or developing chemical munitions) are China, Vietnam, Taiwan, North Korea, Myanmar/Burma, Ethiopia, Egypt, Libya, Israel, Iran and Syria. Iraq was one state that openly declared the existence of its chemical weapons though its stockpile is presently being destroyed by the United Nations Special Commission on Iraq (UNSCOM). The others refuse to make such admissions.

While much is now known about Baghdad's chemical arsenal, there are large gaps in information concerning other Third World proliferators. Most of them probably possess blistering agents such as mustard gas and some may have nerve agents as well. In some instances blood agents and choking agents may also be in the process of development. North Korea, Iran, Syria and Israel also probably have some form of missile delivery system for chemical weapons, while Egypt and Libya have shown interest in such a capability. (Iraq's capability appears to have been destroyed.) States that probably have chemical weapons or a chemical weapons programme (those which US government officials mention off the record, and where there is less certainty) include Laos, Indonesia, South Korea, Thailand, Angola, Argentina, South Africa, Cuba and Pakistan. A third group consisting of those states whose foreign and domestic adversaries claim that they possess chemical weapons but where evidence is often scanty includes Afghanistan, Chad, Chile, El Salvador, Guatemala, Jordan, Peru, Mozambique, Nicaragua, Sudan and the Philippines.[1]

Significantly, of the first group, North Korea, Egypt, Libya, Iraq and Syria have not signed the Chemical Weapons Convention (CWC). Of course these states represent a very small group among the suspected proliferators, but it is their programmes which have always been of the greatest concern. After all, it is these states which are located in regions – the Middle East and North Asia – where the impact of proliferation on regional security may be the most adverse. Furthermore there is every indication that some of these states are intensifying their proliferation efforts. In testimony before the Senate Governmental Affairs Committee in February 1993 on 'Proliferation Threats of the 1990s', the Director of the Central Intelligence Agency, James Woolsey, stated that Libya, which already has the largest chemical weapons production plant in the world, was constructing a second chemical weapons production facility to function alongside the original Rabta facility.[2] Other evidence tends to indicate that the Syrians are also continuing to refine and expand their chemical missile warhead capability while the Iraqi programme could be resuscitated once outside controls are eased.

Furthermore, it would be unwise in all instances to equate signature of the CWC with an unconditional readiness to surrender chemical weapons. Iran, for example, has signed the CWC but has given no indication of halting its chemical weapons programmes. In this regard Director Woolsey commented on 24 February 1993 that Iran 'has an active chemical warfare programme ... [producing] primarily choking and blister agents ... [and] may also have a stockpile of nerve agents'.[3]

Information concerning the scale and scope of biological weapons proliferation is extremely difficult to come by and estimates of who has acquired such systems and who is developing biological weapons can vary. Problems of identification and detection are more challenging than with chemical weapons because of the small size of the research and production facilities. Thus in 1988 CIA Director William Webster stated: 'The moral barrier to biological warfare has been breached. At least 10 countries are working to produce biological weapons.'[4] That same year the US State Department concurred with Webster's figures[5] but in March 1991 USN Director of Naval Intelligence Rear Admiral Thomas A. Brooks informed a Congressional subcommittee that the United States had identified at least seven countries as biological weapons possessors or biological weapons-capable.[6] Third World countries which are sometimes mentioned as possessing or moving towards production include Iraq, Iran, North Korea, Egypt, Israel, Syria and Libya, some of which have not signed the Biological Weapons and Toxins Convention, while others which have signed it have yet to show that they are uninterested in producing such munitions.

While most states do not like talking about their chemical and biological programmes, they sometimes expose their true intentions. In the early 1970s Egyptian President Anwar Sadat stated that 'we have the instruments of biological warfare in the refrigerators'.[7] In October 1988 Hojat ol Eslam Akbar Rafsanjani, then commander-in-chief of the Iranian armed forces and speaker of parliament, was unequivocal:

> With regard to chemical, bacteriological, and radiological weapons training, it was made very clear during the war that these weapons are very decisive. It was also made clear that the moral teachings of the world are not very effective when war reaches a serious stage and the world does not respect its own resolutions and closes its eyes to the violations and all the aggressions which are committed in the battlefield.... [Consequently] we should fully equip ourselves both in the offensive and defensive use of chemical, bacteriological and radiological weapons.[8]

While Iranian officials have since become much more circumspect with regard to their non-conventional weaponry programmes, there is as yet no evidence that the Iranians have actually shifted away from Rafsanjani's explicit policy pronouncement. CIA Director Woolsey stated in March 1993 that Iranian biological weapons 'if not already in production, probably are not far behind'.[9]

The Iraqis are the only Third World state to have admitted to possessing chemical weapons, but with regard to biological weaponry they have not been so open. However in November 1990, PLO Chairman Yassir Arafat announced after a meeting with Saddam Hussein that should war break out, Iraq would employ anthrax.[10] Significantly, CIA Director Woolsey stated in March 1993: 'Iraq's biological weapons capability is perhaps of greatest concern. Baghdad had an advanced programme before Desert Storm, and neither war nor inspections have seriously degraded this capability.'[11] With Iraq's programme still extant it is quite improbable that Iran would be willing to curtail its own efforts in this field, while Israel will be forced to keep its various options open.

By the year 2000 it is estimated that there will be over 24 states with ballistic missile capabilities.[12] Developing states here include India, Pakistan, Iran, Iraq, Egypt, Syria, Israel, Libya, Saudi Arabia, North Korea, Taiwan, South Korea, Afghanistan, Vietnam, Yemen and Algeria. Most of the missiles presently found in Third World countries have been purchased but a large number of these states are now engaged in production of their own models. Soon, many Third World missile forces – some of which are already quite substantial – will be complemented by more and more sophisticated home-manufactured systems.

Important missile development programmes now under way include: Pakistan's Hatf missile family (with ranges of 80, 100, 300 and 600 km); India's Rohini (150 km), Privthi variants (150 and 250 km), Agni (2,500 km), and ASLV space launcher (4,000 km); Brazil's MB/EE (150 km), SS variants (300 and 600 km); Iran's Mushak (200 km), Scud B domestic variant (300 km), Nodong 1/M18 and associated variants (1,000 km); Egypt's Project T (450 km), Vector (600 km); North Korea's Nodong variants (1,000 km+); Taiwan's Tien Ma (950 km); and Libya's al Fatah (950 km). The state of Iraq's residual missile production programme is unclear though it now appears that the bombing of 1991 did not destroy it totally.

What is evident from the list of missile-deploying and developing states is that there is a strong linkage between those countries deploying and developing ballistic missiles and those countries seeking to acquire chemical and biological weaponry (as well as nuclear weapons). One can go further and argue that advanced ballistic missile programmes are often matched by relatively good progress in non-conventional weapons production. It is quite possible that a number of Third World countries view their ballistic missiles as means for delivering non-conventional ordnance. This makes sense because the accuracy of many of these missiles (deployed or in development) is not of a sufficient standard for the targeting of military objectives using conventional warheads. In addition, by configuring missiles to deliver non-conventional warheads against large population centres, weapons of terror (through the arbitrary delivery of small high-explosive payloads) may be transformed into potential weapons of mass destruction. Iran, Iraq, Syria, North Korea, Israel and perhaps Egypt are believed to have succeeded in manufacturing chemical warheads for their missiles. Israel is believed to have a nuclear missile warhead technology.

Even more than in the field of chemical weapons proliferation, a number of Third World countries in the Middle East and South and North Asia are intent upon continuing with their missile development programmes. The end of the Cold War has certainly not blunted their incentives to acquire such systems.

The end of the Cold War and continuing demand

That the decline of superpower tensions has not led to an across-the-board decrease in incentives or efforts to proliferate chemical and biological ordnance and ballistic missile delivery systems should not be surprising. While it was in the context of superpower competition that the Third World proliferation process gained momentum, that competition also helped to keep proliferation to some extent in check. The assurance of superpower political, financial and military (conventional and non-conventional) support moderated the pressure to acquire these costly and controversial systems. This was clearly the case in the field of nuclear proliferation but it is reasonable to assume that the impact was felt with regard to other weapons of mass destruction and delivery system projects as well.

South Korea illustrates the point. During the Cold War, under US protection, it did not press forward with missiles and non-conventional weapons, despite its hostile environment. But a US withdrawal from Asia would press South Korea towards acquiring new weapons systems of its own.

Already, there are examples of isolated countries feeling forced to take the non-conventional and missile routes. Both Israel and, until recently, South Africa felt extremely threatened and neither believed that it could rely on other states to protect its vital interests. Isolated and beleaguered, they turned to weapons of mass destruction and the development of various means of delivery. Had either believed that it could rely on allies for its protection, it is possible that the projects involved would have been followed with less intensity and dedication.

Thus in the 1990s a key factor associated with potential proliferation – certainly in the North Asian arena – is the continued willingness of the United States to extend its security net over a number of countries which feel increasingly insecure. If, for example, South Korea does not feel able to rely on US promises of support, the pressures to expand non-conventional weaponry and ballistic missile programmes may become overwhelming.

In the Middle East it is the collapse of Soviet power that impacts most directly on incentives to proliferate. Syrian interest in upgrading its missile and non-conventional forces has undoubtedly been reinforced by the diminution of Soviet might and the perceived unwillingness of Russia to support it financially, politically and militarily to the degree it has in the past.

Additionally, in many cases the proliferation process has only been partly related to superpower challenges. Most developing-country proliferators have viewed their missile, chemical and biological forces in intra-Third World terms – that is, dedicated to threats deriving from other developing countries – and thus the demise of the Soviet Union has had little impact on their rationales. If anything, the more unstable and unpredictable environment that has emerged, in which Third World enemy countries may feel less restrained by the superpowers, strengthens arguments for acquiring weapons of mass destruction and ballistic missiles.

There is certainly historical evidence that Third World chemical weapon and ballistic missile arsenals have usually been viewed as means of countering Third World power. Thus in none of the three authenticated uses of chemical weapons since the Second World War were such weapons used against a superpower (the cases being those of.Egypt in Yemen, 1963–8; Iraq in the Iran–Iraq war, 1983–8; and Iraqi forces in Kurdistan, 1987–8). Likewise in none of the approximately 60 instances of alleged chemical weapons use were superpowers or industrialized countries the targets. Of the five cases of missile employment since 1945 three have been intra-Third

World (the Yom Kippur war of 1973, the Iran–Iraq war and the war in Afghanistan during the 1980s and early 1990s). There have been no cases of biological weapons employment.

There have been allegations of US, European and Soviet chemical weapons employment in the Third World (for example in Nicaragua, Guinea-Bissau and Afghanistan) and Third World missiles have been employed against a NATO base (Libyan missiles against Lampedusa in 1988) and against US/allied forces in the Gulf in 1991. Yet these instances are of relatively small significance compared with the cases of intra-Third World use. In the cases of the chemical weapons allegations, most are probably ill-founded, while in the cases of the missiles the 40 or so used against the US and its allies have to be compared with the thousands employed by developing countries against other developing countries. Scenarios involving the industrialized world have thus traditionally not dominated Third World acquisition rationales.

Of course one of the reasons why Third World countries have refrained from employing chemicals against the industrialized world is that they prefer to employ their weapons against parties who have no ability to respond in kind (and no equivalent capability to impose large-scale destruction). This has long been the case and has served to reinforce the desire of many developing states to acquire such systems as vital means of deterrence. The fear of being caught empty-handed and unable to respond in kind has survived the Cold War and continues to motivate Third World chemical weapons proliferation. An arms race imperative also appears to be at work in the area of biological weapons research and ballistic missiles, though in the latter case there is little evidence that possession of missiles by two countries results in mutual deterrence.

In the final analysis proliferators in the post-Cold War world continue to be attracted by chemical/biological weapons and ballistic missiles on account of their military capabilities, which have at the very least remained constant and in some instances have improved; more importantly, the perception of their political utility has arguably also increased.

Demand and political/military utility

The United Nations first classed chemical weapons alongside biological and atomic weapons, characterizing them as weapons of mass destruction. Since then, many have come to view chemical weapons as the 'poor man's

A-bomb'. There is some truth to this description. Clearly a number of developing states have always viewed their chemical weapons systems as a partial (though not totally satisfactory) means of opposing enemy nuclear weapons. The Middle East Arab states are the best examples.

Chemical weapons therefore are certainly not called poor men's A-bombs because all the states producing them are poor. They are of course usually states with modest technological capabilities and this reflects the major attractions of chemical weapons: the relative ease of their production (compared with nuclear weapons) and their cheapness. Chemical weapons in developing states are thus often acquired as a substitute for nuclear weapons. However, that is not the only possibility in principle or practice. They can also be a stopgap measure until nuclear weapons are acquired, and/or a deterrent to other chemical weapons, or a less destructive complement to nuclear weapons.

However, chemical weapons are 'poor' weapons for other reasons which raise questions about their overall rationale. First, it is extremely difficult to assess the amount of damage chemical weapons may inflict: they do not act directly on the individual but on the environment, and are subject to the vagaries of that environment. Thus, according to one atmospheric-dispersion model, 5 tonnes of (Sarin) nerve gas disseminated into the air could lead to 50 per cent casualties over an area of 4 square kilometres, though this figure should be multiplied or divided by a factor of three or four depending upon the weather.[13] Military planners prefer to maximize certainty when preparing for battle and chemical weapons obviously cannot readily help with this task. With nuclear weapons the quantity and pattern of fallout is affected by weather conditions, but this is only one aspect of the effect of nuclear weapons.

Second, it is not only the weather that degrades chemical weapons' performance. The above figures refer to an attack on open countryside, not on an urban target where dispersion would be restricted and casualty ratios would be lower. Third, casualties would be dramatically reduced if the targeted population wore protective gear. Casualties during both the First World War and the Iran–Iraq war dropped considerably once soldiers were fully equipped with gas masks.[14]

Chemical agents do not have a good record as battlefield weapons or as weapons of mass destruction. Theoretically, they certainly do not compare with nuclear weapons. For example, it has been estimated that the 50 per cent casualty contour for an air-burst fission bomb of 10–15 kilotonne yield would be about 30 square kilometres.[15] Fetter demonstrates that while a 20-kilotonne nuclear missile warhead would result in 40,000 deaths if dropped on a large city with an average population density of 30 per hectare, a 300-kg Sarin missile warhead would result in anything between 200 and 3,000 deaths.[16] Chemical weapons appear nearer to conventional than to nuclear explosions.

Yet these considerations continue to have only a marginal impact upon acquisition incentives. Third World countries, after all, are still well aware of the instances where chemical agents have proved devastating. In the civil accident at Bhopal in December 1984, 30 tonnes of methyl iscoyanate vaporized under favourable weather conditions in an extremely densely populated area. Here the 50 per cent casualty contour extended to an area of approximately 70 square kilometres. More recently, at Halabja, the Iraqi army was able to cause thousands of casualties among the helpless and undefended Kurdish population through the employment of chemical weapons. The point here is that, while under certain circumstances chemical weapons may not be very effective at causing casualties, in other circumstances they may be devastating. Military planners, as we know, sometimes act on the basis of the 'best case' rather than the 'worst'.

Uncertainty may be an advantage to the employer of chemical weapons. A prospective target population can never be totally sure what the effects of such weapons may be and they will wish to prepare for the 'worst case'. Related to this is the psychological dimension. Chemical weapons have a unique capability for inflicting terror among target populations. Thus, while chemical weapons may not match the destructive effects of nuclear weapons, they can have a major deterrent and force multiplier effect. A number of developing countries which recognize this remain attracted to chemical weapons.

There is great concern that biological weapons may emerge as true weapons of mass destruction whose effects may be comparable to, and may even surpass, those of small nuclear explosions. Fetter's calculations show that a missile warhead consisting of 30 kg of anthrax spores, if employed against a large city with a population of 30 per hectare, could kill between 20,000 and 80,000.[17] In some scenarios it is possible that 30 kg of anthrax spores could cause double the casualties of a 20-kilotonne nuclear warhead.

Yet there is little experience of use of biological weapons, and their effects are at least as unpredictable as those of chemical weapons. It has therefore been argued by a number of analysts that biological weapons, while theoretically devastating, have limited military utility. They may take days or even weeks to infect their target population who may in some cases be inoculated against the disease. Friendly troops will also have to be inoculated (if the disease chosen is a transmissible one) and this may be time-consuming and give away any element of surprise. In addition, there are important challenges to be faced concerning storage and maintenance, the fragility of micro-organisms in the atmosphere, dependence upon environment and difficulties in handling. All these factors may severely tax Third World proliferators.

On the other hand, the merits of biological weapons for Third World proliferators may well outweigh these disadvantages. According to one analyst the advantages of biological warfare include 'the extreme potency [of biological agents] so that small quantities are necessary; the lack of signature if an endemic disease agent is selected; the ease of covert production using dual-purpose facilities; lack of collateral damage; lack of risk if a non-transmissible agent is selected; and large-scale/strategic small-scale/tactical use or covert use'.[18]

Significantly, recent developments in biotechnology may now be acting to decrease the inherent drawbacks of biological munitions and to increase their military utility. It is possible that types are being manufactured against which there is no known defence. These factors could well increase the demand for biological weaponry over the next decade. It is therefore possible to argue that concern over the proliferation of nuclear weapons should be matched by concern about biological weaponry.

Ballistic missiles continue to remain attractive to a number of Third World countries because they provide these states with valued military capabilities. First, and perhaps most importantly, they provide the possessor with a means of penetrating enemy air defence which would otherwise be difficult for some Third World countries. The most obvious example is Syria, whose air force is grossly inferior to that of Israel. Missiles provide Syria with the one virtually guaranteed means of penetrating Israeli air space and striking at military and civilian targets.

The implications of virtually guaranteed penetration extend well beyond intra-Third World terms. Few developing countries have the capability of penetrating the air defences of developed states, whether such defences be over such states' homeland or over their projected forces. Missiles could be crucial here and it is possible that a number of Third World countries, fearful of future industrialized world power projection, may be increasingly attracted to this rationale. Also appealing is the speed with which missiles can reach their targets, especially when surprise may be vital to specific operations.

Missiles also widen the circumference of operations of developing states and enable them to target or even to attack parts of the world their other weapons cannot reach. In the Middle East good examples of this are Iraq and Iran against Israel. Colonel Gaddafi of Libya has also expressed his desire to acquire a weapon capable of hitting New York,[19] as has the terrorist Abba Abbas.[20] In North Asia ballistic missiles will enable North Korea to target states it otherwise would have great difficulty in reaching.

An important attraction for Third World states is that ballistic systems can be configured to deliver a wide variety of ordnance including non-conventional munitions. Indeed, it can be argued that, given the inaccuracy of presently deployed Third World missiles (and the fact that it is quite unlikely that accuracy will be dramatically improved in the near future), these weapons only make sense with non-conventional warheads. As noted above, numbers of Third World countries continue to invest large amounts of effort and resources in developing such technology. Also, ballistic missiles have capabilities for causing terror among civilian populations in a manner which aircraft often do not. This is a capability that has been demonstrated on a number of occasions including the Second World War, the Iran–Iraq war and, more recently, the 1991 Gulf war (see below). The speed of the weapon, the fact that in many states there is no technology available for warning, and the arbitrary nature of the missile's final destination, all serve to reinforce this terror aspect.

Lessons of the Gulf war and continued demand

It can be argued that for at least some states – especially in the Middle East – events in the early 1990s actually increased incentives to acquire chemical/biological weapons and ballistic missiles, and the Gulf conflict in 1990–91 particularly demonstrated the military and political utility of these systems. Saddam's much-trumpeted non-conventional arsenal of course did little militarily to help him during his disastrous war against the US-led coalition. Despite the enormous amount of money and resources that went into his non-conventional forces, they appeared to have no impact on the battlefield. As a consequence, it could be argued that such weapons were not worth the effort required to deploy them. Since the weapons neither deterred nor were readily employable, this conflict should therefore surely serve to undermine incentives to proliferate.

Yet, non-conventional weapons had an impact on the causes and conduct of operations which underlined their utility.[21] The first major point to note is that, while Saddam had hundreds of aircraft and thousands of tanks, it was his chemical weaponry that caused particular concern among allied governments and publics alike. Following the war George Bush stated that one of his greatest fears was that Saddam would resort to chemical weapons and that all planning was predicated upon this assumption.[22] Certainly Saddam had been warned that his use of such weaponry would be regarded as a significant escalation, different from the employment of conventional weaponry. In January 1991 US Secretary Baker handed a letter from Bush to Iraq's Foreign Minister Tariq Aziz (he refused to accept it but its contents were made public). This stated that

> The United States will not tolerate the use of chemical or biological weapons, support of any kind of terrorist actions, or the destruction of Kuwait's oilfields and installations. The American people would demand the strongest possible response. You and your country will pay a terrible price if you order unconscionable actions of this sort.[23]

This concern about non-conventional weaponry was not limited to governments but was strong among allied publics as well. A debate ensued in the media over the scale of Saddam's non-conventional programme, the likelihood that he would resort to chemical weapons employment and the probable effect of such munitions on allied troops. The subject of this intense and uncomfortable debate gave the impending war an apocalyptic aura which helped to cover the true discrepancy of forces. Undoubtedly, the image of soldiers wearing gas suits and masks became the central image of the conflict in the period prior to fighting. It was this concern among the public and fear that Iraq's non-conventional programme would improve that helped Bush argue the 'fight now or pay later' rationale for the war.

The preparation and conduct of the war was also influenced by the existence of chemical and biological weapons. Not only did medical measures have to be taken and specific training programmes followed, but bombing plans had to prioritize non-conventional production and stockpile facilities (at the expense of other targets). Even the termination of the war demonstrated that the allies remained concerned about Iraq's non-conventional weapons programme: Resolution 679 demanded that all its non-conventional facilities and stockpiles be surrendered and destroyed and that steps be taken to ensure that the country never developed such weaponry again.

Chemical weapons demonstrated themselves to be ones about which allied countries were, if not fearful, then quite apprehensive. Therefore, potentially, they appeared to be excellent weapons of deterrence. Of course during the Gulf war not only were they not employed but they failed to deter. Yet had Saddam's non-conventional weapons been more sophisticated and more deadly (for example had he been known to deploy effective chemical missile warheads, biological weaponry or nuclear weaponry), it is quite likely that the allied calculation behind the decision to go to war would have been radically affected. Had Saddam waited longer before committing his aggression, allowing his non-conventional weapons programme to mature, the response of the West might have been quite different.

Of course Saddam's non-conventional weaponry also failed to play a decisive role in the conflict because of the mass of forces ranged against him. But what if in a future case an alliance is not so strong, if the case of aggression is not so explicit and the balance of conventional forces is not so great? Then the existence of chemical weaponry could play an important deterrent role and possibly a significant role on the battlefield. Since most Third World proliferators are, as noted, interested in intra-Third World scenarios, this argument is even stronger.[24]

Furthermore, among the explanations why Saddam did not employ his non-conventional weaponry was his fear of massive conventional retaliation (from the allies) and non-conventional strikes (from the Israelis). In the former instance, most Third World states cannot match the conventional power of the allies and as a result could be forced down the non-conventional weapons path. With regard to the latter point, the Israelis certainly believe that their non-conventional arsenal deterred Saddam from employing his own non-conventional weaponry against them.

In other words the failure of non-conventional weaponry in the Gulf war was a failure of Saddam's grand strategy and not a failure of the weapons systems themselves. The fact that a number of countries in the developing world (especially in the Middle East where they had a good view of the war) are continuing to work to acquire more non-conventional weaponry may be a result of this reading of the conflict.

The Gulf war had even more explicit lessons for missile proliferators. Unlike chemical weaponry, ballistic missiles were employed and they were demonstrated to be effective. True, they were inaccurate, hitting very few military targets, and they failed in their major political objective of bringing Israel into the war. Nevertheless they served other important functions.

First, they demonstrated penetration capabilities. The missiles succeeded in penetrating Israel's formidable air defence and striking at urban centres deep inside the country. No Arab air force has that capability and therefore the war underlined the fact that missiles provide states with options they otherwise may not possess. Second, they created a degree of disruption and concern among the Israeli population. The war showed that missiles do possess terror qualities which should not be underestimated. It is very difficult to assess whether the civilian fear was being caused by the missiles themselves or by concern that future missiles would be armed with chemical weapons. Whatever the case, the war pointed to the danger of future missile forces armed with chemical warheads and highlighted their attractions. Third, the greatest single loss of American life was caused by a missile strike – the hit on the army barracks towards the end of the war – and while this was probably an instance of chance rather than correct targeting the incident did point to the dangers that would be encountered should missile accuracy improve considerably. Finally, the war demonstrated that search-and-destroy missions against mobile launchers are demanding of resources and highly ineffective. Probably no, or very few, mobile launchers were destroyed by allied air strikes.[25] In addition, despite some optimistic claims about Patriot interception success rates, it now appears that missile warheads succeeded in penetrating these defences in most instances.

The 1991 Gulf war thus served to reinforce preferences for proliferation (especially in the Middle East) rather than acting as a damper on procurement and development rationales. While many in the West argued that the war demonstrated ballistic missiles to be ineffective weapons, this is not the interpretation made by states in the Middle East (and possibly elsewhere). Israeli pronouncements and policies since the end of the war show quite clearly that Jerusalem believes that one of the major threats facing Israel during the 1990s will be ballistic missile forces. This is despite any 'flaws' rendered apparent by the Gulf war or the strenuous efforts of arms controllers.

While missiles were shown to have severe limitations, these were not beyond the remedy of more and more sophisticated systems. It should not be forgotten that ballistic missiles employed during the 1973 Yom Kippur war proved even less useful militarily than they would prove to be in 1991, yet following that conflict there emerged a major missile build-up in the Middle East. Utility appears to be very much in the eye of the beholder.

Control and continuing demand

Efforts aimed at constraining the proliferation of chemical/biological weapons and ballistic missiles have been only partially successful. When countries are intent upon acquiring these systems then arms control appears to have had only a limited impact, increasing the difficulties, lengthening the period of development programmes and raising costs.

While a country such as India has admitted that the MTCR has hampered its rocket production efforts, few countries have been forced to abandon their rocket production programmes because of supplier regimes. Where there have been instances of missile programmes being surrendered – as in South Africa and Argentina – these have usually been the result of broader political considerations rather than simply because of the control regime itself.[26]

It is most unlikely that a control regime such as the MTCR can prove totally successful in curbing missile proliferation, let alone in weakening Third World demand for the weapons. The regime is perceived by many in the developing world as a discriminatory cartel of industrialized-country interests at variance with the sovereignty and independence of Third World states. Few countries interested in missile systems willingly support the regime either in their trade policies or in terms of halting their domestic production programmes. The regime after all offers Third World states little in terms of positive gain as it does little to address their security concerns directly. It cannot do away with missiles already deployed (and there are thousands of these) nor does it provide alternatives for states which view missiles as providing them with particular military options they otherwise would not have.

So far at least the MTCR has failed to bring all major suppliers into the regime with the result that there is serious 'missile leakage' into the Third World. The key problems here are China and North Korea, which continue to supply missiles and missile technology to countries in South Asia and the Middle East. With these two countries willing to supply advanced technology to regions of great insecurity, it is not unexpected that demand in these theatres continues to be high.

The area of chemical and biological weaponry is somewhat different in that a supplier cartel is accompanied by international treaties outlawing the weapons themselves. However, where demand for the weaponry is very strong – for example in Iran which suffered severely at the hands of Iraqi chemical weapons forces – then signing an agreement may not necessarily mean willingness to abandon production capabilities. A long and difficult road has yet to be travelled before there is real movement towards disarmament and even those who have now committed themselves to disarmament may have many opportunities to change their minds.

In any event, readiness to sign the conventions may be not a reflection of the strictness of control but a result of other political and military considerations. Israel, for example, retains a nuclear capability and is thus not under defence pressure to hold onto chemical capabilities. What Jerusalem seeks is a divorce between these two weapons systems: an effort which Arab states can be expected to resist strongly. Thus most of those states which have so far refused to sign up to the CWC will probably not do so in the foreseeable future.

Where the demand is high, efforts will continue to be made by signatories and non-signatories to circumvent the various non-proliferation regimes. Just as in the area of ballistic missile arms control, control efforts related to chemical weaponry will, in certain circumstances, slow down the programmes by increasing their difficulty and expense. Whether arms control will be able to halt proliferation where demand is strong seems questionable.

If demand is linked to perceived utility and arms control cannot fully stop demand, then perhaps the best focus for effort is to decrease the perceived utility of these systems. Improved anti-chemical and anti-biological defences and ATBMs must surely be part of the answer. At the same time efforts dedicated to reaching political solutions so as to make the weaponry itself redundant should receive priority. This is the approach already being taken in the present Arab–Israeli discussions, in which issues of borders and mutual recognition are taking precedence over arms control.

In the final analysis arms control is a necessary but not a sufficient mechanism for constraining proliferation and countering continued demand. It can slow proliferation down by making it more difficult but it cannot fully neutralize the desire by all states to acquire missiles or non-conventional weaponry. It can be argued that arms control helped to prevent a rush to non-conventional and ballistic missile acquisition and that we are now concerned only about a small group of states predominantly located in the Middle East and to some extent in South and North Asia. Yet it is this small number of states whose proliferation policies continue to pose severe challenges to regional and international security.

Consequently, both Western and Japanese foreign policies must be dedicated towards complementing supplier restraints with efforts aimed at convincing and assuring procurers that their security interests are best served by cutting back rather than building up weapons of mass destruction. If the demand side is ignored, then supplier restraints will increasingly be undermined and their objectives frustrated.

Notes

1 Elisa D. Harris, 'Chemical Weapons Proliferation: Current Capabilities and Prospects for Control', Appendix 2, *New Threats: Responding to the Proliferation of Nuclear, Chemical and Delivery Capabilities in the Third World* (Lanham, MD: Aspen Strategy Group and United Press of America, 1990).

2 *Arms Sales Monitor*, No. 19, 15 March 1993.

3 *Chemical Weapons Convention Bulletin*, 19 March 1993, p. 21.

4 *Washington Times*, 13 December 1988.

5 Statement by Thomas J. Welch, deputy assistant to the Secretary of Defense to Subcommittee on Arms Control, International Security and Science of the Committee on Foreign Affairs, 3 May 1988.

6 Statement of Rear Admiral Thomas A. Brooks, USN Director of Naval Intelligence, before the Seapower, Strategic and Critical Materials Subcommittee of the House Armed Services Committee on Intelligence Issues, Washington, DC, 7 March 1991.

7 Reported in *Foreign Broadcasting Information Service*, 2 September 1988.

8 Reported in *Foreign Broadcasting Information Service*, Near East and South Asia, 7 October 1988.

9 *Arms Sales Monitor*, No. 19, 15 March 1993.

10 *International Herald Tribune*, 6 November 1990.

11 *Arms Sales Monitor*, No. 19, 15 March 1993.

12 Keith B. Payne,'Defence Against Missile Proliferation', *Jane's Intelligence Review*, May 1992.

13 Julian Perry Robinson, 'Chemical Weapons Proliferation in the Middle East', Institute for East–West Security Studies, New York, 7–9 February 1990.

14 T. McNaugher, 'Ballistic Missiles and Chemical Weapons: The Legacy of

the Iran–Iraq War', *International Security*, Vol. 15, No. 2, Fall 1990, pp. 5–34.

15 Robinson, op. cit.

16 Steve Fetter, 'Ballistic Missiles and Weapons of Mass Destruction: What is the Threat and What Should be Done', *International Security*, Vol. 16, No. 1, Summer 1991.

17 Ibid.

18 Graham S. Pearson, 'Biological Weapons: Their Nature and Control', in E. Karsh, M. Navias and P. Sabin, *Non Conventional Weapons Proliferation in the Middle East* (Oxford: Clarendon Press, 1993), p. 113.

19 Quoted in L. Spector, *Nuclear Ambitions* (Boulder, CO: Westview Press, 1990), p. 371.

20 'A Terrorist Talks about Life, Warns of More Deaths', *The Wall Street Journal*, 10 September 1990.

21 These arguments are taken from M. Navias, 'Saddam's Scud War and Ballistic Missile Proliferation', *CDS Papers*, No. 6, 1991.

22 H. Sidey, 'Twas a Famous Victory', *Time*, 6 January 1992.

23 *International Herald Tribune*, 14 January 1991.

24 One may also argue that the Gulf war showed the dangers of power projection for the Third World. While such scenarios may not have been very important until now in driving demand, they may become more important.

25 See Navias, op. cit.

26 For discussions of the MTCR, see J. E. Nolan, *Trappings of Power: Ballistic Missiles in the Third World* (Washington, DC: Brookings, 1991), and M. Navias, 'Ballistic Missile Proliferation in the Third World', *Adelphi Papers*, No. 252, London, IISS, 1990.

CHAPTER 4
THE NUCLEAR NON-PROLIFERATION REGIME BEYOND 1995

MITSURU KUROSAWA

With the end of the Cold War, the international security environment is rapidly changing. With East–West tensions diminished, the possibility of an all-out nuclear exchange between the United States and Russia is substantially reduced. However, the international community cannot afford to become complacent because the threat of proliferation of weapons of mass destruction and their means of delivery has received a new impetus in spite of the end of the strategic nuclear arms race between the superpowers or perhaps in its place.[1]

The revelations about Iraq's clandestine nuclear weapons programme, the collapse of the Soviet Union and the emergence of new independent states with nuclear weapons on their territories, and the announcement of an intention to withdraw (later rescinded) from the Nuclear Non-Proliferation Treaty (NPT) by the People's Democratic Republic of Korea, all prompted the international community to respond with effective measures.

In 1995, a conference of signatories will be held to review the NPT. Its purpose will be to decide by majority vote whether the treaty shall continue in force indefinitely, or shall be extended for an additional fixed period or periods.

I would like to examine first the structure and characteristics of the various elements in the nuclear non-proliferation regime, second the 1995 NPT review and extension conference, third the strengthening of the nuclear non-proliferation regime, and finally the issue of nuclear export control.

The nuclear non-proliferation regime

The NPT

The Nuclear Non-Proliferation Treaty is the foundation of the nuclear non-proliferation regime and has served as the principal legal barrier to nuclear weapons proliferation for more than 20 years. In the 1960s, many, including President Kennedy, had predicted that 20-25 states would soon possess nuclear weapons. From that perspective the NPT has succeeded in preventing the spread of nuclear weapons since there are still only five accepted nuclear weapons states and the treaty now has more than 155 parties.

In recent years, France, China and South Africa, among others, have acceded to the treaty, making it more universal. However, there are some states that still wish to stay outside its strictures, and these include Israel, India and Pakistan.

Iraq, a party to the treaty since 1969, has violated its obligations by clandestinely developing nuclear weapons. North Korea, which became a party to the treaty in 1985 and accepted the International Atomic Energy Agency (IAEA) safeguards as late as 1992, announced its intention to withdraw from the treaty in the face of a demand for a special inspection by the IAEA. Although it has since renounced this intention under outside pressure, suspicion still remains as to its real ambitions and actions. Iran and Libya, which are also parties, are suspected of trying to develop nuclear weapons. These cases show that even if a state is a party, we may not be able to trust it and there is a need for exacting intelligence. As the principal objective of the treaty is to prevent the emergence of new nuclear weapons states (NWSs), every activity connected with the development of nuclear weapons by non-nuclear weapons states (NNWSs) is prohibited, while NWSs are free to improve their nuclear weapons capabilities. Thus, the treaty divides nations into two categories with different rights and obligations. The most serious criticism of the treaty concerns this discriminatory element although the fact that more than 150 NNWSs have nevertheless acceded to the NPT so far means that they have found merit in it.

A provision inserted to mitigate this criticism is Article VI, which stipulates that NWSs must pursue negotiations in good faith on nuclear disarmament measures. Many NNWSs thought that non-proliferation was a necessary first step for the NWSs to begin negotiations on nuclear disarmament. The announcement by President Johnson of the beginning of negotiations with the USSR on the limitation and reduction of strategic nuclear arms was made on the same day that the NPT was opened for signature, in order to emphasize the connection between the two issues.

NWFZs

A second approach to preventing the spread of nuclear weapons is the establishment of nuclear weapon-free zones (NWFZs) on a regional basis. NWFZs are regional measures to complement the NPT taken at the initiative of the states directly involved. Article VII of the NPT encourages the creation of NWFZs.

A treaty establishing an NWFZ includes a commitment to non-deployment and places an obligation on NWSs to respect the status of an NWFZ and not to use or threaten to use nuclear weapons against states within the zone. The first treaty that established an NWFZ was the 1967 Treaty of Tlatelolco in Latin America, but it has long suffered from the absence of the important states like Argentina and Brazil. However, Argentina and Brazil, which recently agreed to accept IAEA safeguards, have decided to accept the treaty. Chile will do so too, and Cuba has indicated interest in joining the treaty as well.

The main reason why Argentina and Brazil have not joined the NPT is its discriminatory character. However, NWFZs are less discriminatory, and moreover the obligation not to use nuclear weapons is included in the concept of an NWFZ but not within the concept of the NPT. All five NWSs have ratified the Additional Protocol II of the treaty, which prohibits the use of nuclear weapons against the parties constituting an NWFZ.

A second treaty is the South Pacific Nuclear Free Zone Treaty of 1985. This also includes Protocol II, which prohibits use of nuclear weapons against member parties. Only two NWSs have ratified this protocol so far.

IAEA safeguards

A third element in the nuclear non-proliferation regime is the conclusion of agreements with the IAEA for the application of its safeguards system. Parties to the NPT or to a treaty establishing an NWFZ are under the obligation to accept safeguards for all their peaceful nuclear activities, but as the recent cases of Iraq and North Korea have shown, some parties fail to declare all facilities possessing nuclear material. The right of special inspection by the IAEA, which is thereby entitled to demand access to a site or material that is not declared by the parties, was reconfirmed by the Board of Governors of the IAEA in February 1992.

In the case of non-parties, according to the traditional interpretation, the IAEA safeguards have been applied only to nuclear material or facilities that the non-party has received from parties. Some states like Canada, Australia and the US have applied full-scope safeguards, that is safeguards that cover all nuclear materials in peaceful uses of nuclear energy in a country, in accordance with their domestic laws or policies, but many other suppliers have not, although it has been generally recognized that in order to prevent proliferation full-scope safeguards are indispensable.

In April 1992, the 27 nations of the London Nuclear Suppliers Group agreed not to export nuclear-related material or facilities to a nation that has not accepted full-scope safeguards to nuclear material in its present and future activities.

Export control

A fourth and final factor in non-proliferation is the common export policies among nuclear supplier nations. From the mid-1970s, the London Nuclear Suppliers Group, which originally included seven industrial nations and now includes 27 countries, has been working to adopt guidelines to prevent the export of sensitive material and facilities.

In the face of the revelation of Iraq's nuclear weapons development programme following the Gulf war, it was recognized that the traditional approach would not be able to stop a nation like Iraq from going nuclear. In March 1992, the 27 members of the Nuclear Suppliers Group agreed on a new multilateral arrangement for controlling a wide range of nuclear-related dual-use commodities.

The nuclear non-proliferation regime is composed of these international legal agreements and common policies. It is a multifaceted complex, and inevitably includes conflicts between NWSs and NNWSs and between the North and the South.

The NPT review and extension conference

Review and extension

Article X, paragraph 2 of the NPT stipulates the following:

> Twenty-five years after the entry into force of the treaty, a conference shall be convened to decide whether the treaty shall continue in force indefinitely, or shall be extended for an additional fixed period or periods. This decision shall be taken by a majority of the parties to the treaty.

Every five years after the treaty's entry into force, a review conference has been held, that is in 1975, 1980, 1985 and 1990. The fifth review conference is to be held in 1995 in accordance with Article IX, paragraph 3. In December 1992, the UN General Assembly took note of the decision of the parties to the Nuclear Non-Proliferation Treaty to form a Preparatory Committee for a conference to review the operation of the treaty and to decide on its extension. It was decided that the review conference and the extension conference would be held at one and the same time, in April and May 1995 in New York.

With the end of the Cold War, the coming review conference will be held under completely different conditions from the four previous reviews. As the Intermediate-Range Nuclear Forces (INF) Treaty has been completely implemented, and the United States and the Russian Federation have agreed to a deep reduction of strategic offensive nuclear arms, both countries can be said to be proceeding more quickly towards nuclear disarmament. Global strategic circumstances have changed radically with the dissolution of the Soviet Union, the dismantling of the Warsaw Treaty Organization, and the conclusion of the Treaty on Conventional Forces in Europe (CFE), as well as other developments.

Because the conference will deal with both the review and the extension of the treaty, the review conclusions will be critical for addressing the problem of extension. If the treaty's operation is deemed to be excellent, we can expect an indefinite or long-term extension. However, if the result of the examination is negative, the future of the treaty will be controversial.

An extension will be decided by a majority of the parties and because the NNWSs have many more votes, the NWSs in particular have a great many tasks to perform prior to the review conference in order to obtain a decision for an indefinite or long-term extension. These issues are examined below in the section on strengthening the nuclear non-proliferation regime.

The extension conference[2]

The negotiating history of the treaty shows that the provision for an extension conference reflects the desires of the industrial non-nuclear weapons states, since the US draft treaty in 1965, the Soviet draft treaty in 1965, and the US–Soviet identical draft treaty in 1967 all provided for the treaty to be of unlimited duration. There were many states that opposed a treaty of unlimited duration on the grounds that they could not foresee future situations brought on by the rapid development of nuclear science and its technical, economic, political and military implications. Italy submitted an amendment to the draft treaty that said: 'The present treaty shall have a duration of X years. It shall be automatically extended for terms equal to its initial duration for those governments which, subject to six months' notice, shall not have made known their intention to withdraw.'

Switzerland's memorandum also stated: 'It would be preferable that the treaty should be concluded for a definite period, at the end of which a review conference would decide about its renewal. During that interval the NWSs could adopt specific measures aimed at a limitation of armaments. The NNWSs certainly cannot take the responsibility of tying their hands indefinitely if the NWSs fail to arrive at positive results in that direction.'

The NNWSs in the NATO alliance, in particular West Germany, opposed a treaty of unlimited duration. Three governments which preferred a treaty of unlimited duration had to yield to those opposed to it, and the revised draft treaty of January 1968 included the present formula.

The provision gives the parties the opportunity to examine all aspects of implementation of the treaty after 25 years, taking into account the situation at that time and the number and quality of the parties. The unique formula for the decision on an extension, which is to be taken by a majority of the parties, is quite different from the formula for amendment of the treaty, which requires a unanimous affirmative vote by all NWSs and by the members of the IAEA Board of Governors. In the extension process, every party has equal status.

What to decide?

According to the treaty's provisions, there are three possible decisions:

(1) that the treaty shall continue in force indefinitely;
(2) that the treaty shall be extended for an additional fixed period;
(3) that the treaty shall be extended for additional fixed periods.

The third option may mean providing either for plural extension periods that shall be adopted in accordance with given situations, or for automatic extensions after a certain period. Given its text and accepted interpretation, there seems to be no possibility of terminating the treaty at the conference. Its duration is not specified as 25 years. The treaty presupposes a duration longer than 25 years by means of a decision on the additional period at the conference.

Although there will be no problem if the decision is taken to extend the treaty indefinitely, or for an additional period or periods, there still remains the possibility that no decision will be taken; that is, that no proposal will receive a majority vote. In this case, one scholar has argued, 'If a conference is not held or if no agreement is reached, the treaty will continue in force indefinitely until the parties can agree otherwise.'[3] This argument is a reasonable one, but if the proposal that the treaty be extended indefinitely does not receive a majority of votes and no decision is taken, it will be hard to maintain, as he does, that an extension should occur automatically. If no decision is taken, a possibility on which the treaty provides no guidance, the parties will have to resolve the issue at the conference.

In any case, it would be desirable to decide the extension by consensus rather than by majority vote. Measures should be taken prior to the conference to ensure that many NNWSs will agree to an indefinite or long-term extension of the NPT.

Strengthening the nuclear non-proliferation regime

Six measures should be taken to strengthen the nuclear non-proliferation regime while eliminating or mitigating its present discriminatory character.[4] Although some scholars have argued recently for a division between states that support the non-proliferation regime and those that oppose it, instead of the division between NWSs and NNWSs,[5] the latter is fundamental in the analysis of the review and extension conference.

Implementation of the START treaties

START I was signed by the United States and the Soviet Union in July 1991 after nine years of negotiation. Although it has been claimed that this treaty has produced a 50 per cent reduction within seven years, it will in fact reduce the two countries' respective deployed strategic nuclear warheads by one-third, using complex counting rules as regards warheads on specific delivery systems.

Because the Soviet Union itself broke up before ratification, a Protocol to the treaty was signed in Lisbon in May 1992 by the United States, the Russian Federation, Belarus, Kazakhstan and Ukraine. The protocol recognizes the four republics as successor states of the former Soviet Union in order to implement the treaty provisions.

Ratification of the protocol by all five states is necessary for START I to enter into force, but Ukraine has not ratified it so far. Ukraine is demanding security assurances, especially from Russia, the United States and the United Kingdom, and economic assistance for nuclear disarmament. As the consent of Ukraine is indispensable for the implementation of START I, these states should give it security assurances, including negative security assurances by NWSs and positive security assurances that are an improvement on Security Council Resolution 255 (1968); moreover, these states and others should assist in the economic field, and especially in the destruction (dismantling) of nuclear weapons systems.

In January 1993, START II was signed between the United States and the Russian Federation, based on the June 1992 agreement between Presidents Bush and Yeltsin. The treaty will reduce their strategic offensive nuclear arms by two-thirds compared with present levels and will totally eliminate MIRVed ICBMs by the year 2003.

As this treaty will not enter into force unless START I also enters into force, ratification of the protocol by Ukraine is also urgent for START II to operate. Moreover there is some concern that START II will not be ratified by Russia because the leadership of President Yeltsin will not be firm enough and because of powerful opposition to the treaty's content, which some allege to be disadvantageous to Russia.

Tight control of tactical nuclear weapons

After the attempted *coup d'état* in the Soviet Union, President Bush in September and President Gorbachev in October of 1991 announced unilat-

RIDING HOUSE STREET LIBRARY
CITY OF WESTMINSTER
37-49 Riding House St., W1P 7PT

eral measures for tactical nuclear disarmament. They decided to withdraw all land-based short-range nuclear forces, which were stationed mainly in areas where the East and the West were in direct confrontation, and also to withdraw tactical nuclear weapons from surface ships and attacksubmarines.

These unilateral but parallel measures were taken by both countries because there was no need for the deployments after the end of the Cold War and because there was a danger that these tactical nuclear weapons might be obtained by other states or terrorist groups.

The weapons were reportedly all withdrawn to the United States or Russia by summer 1992. Some of them are being destroyed, while others are stored in central facilities. Confirmation that the withdrawal and destruction have in fact taken place is necessary to guard against possible seizures by unauthorized parties. Tight control over these tactical nuclear weapons systems is essential.

Cut-off of nuclear materials production

Neither the INF Treaty nor the START Treaties deal with the destruction of nuclear warheads. The treaties demand only the destruction of delivery vehicles, that is launchers, bombers or missiles. In order to achieve real nuclear disarmament, nuclear weapons and the fissionable materials contained within them must be eliminated. This means that the United States and Russia must stop production of fissionable materials for weapons as a first step. As the United States has already stopped producing nuclear materials for weapons purposes and Russia is planning to do the same, a *de facto* cut-off in both countries is approaching. This measure necessitates the closure of production facilities for military fissionable materials and the placement of all peaceful nuclear activities under the safeguards of the IAEA. In order to prevent fissionable materials recovered from dismantled missiles from being reused for weapons, IAEA safeguards over all recovered fissionable materials are necessary.

In line with the production cut-off by the United States and the Russian Federation, the United Kingdom, France and China should fulfil the same obligation and stop making nuclear materials for weapons purposes. This measure is a prerequisite for persuading Israel, India, Pakistan, North Korea and other states not to increase their nuclear materials for weapons. In order to implement this measure, there must be a substantial increase in IAEA inspectors. Nuclear scientists and engineers who have lost their jobs, notably from the former Soviet Union, could serve as IAEA inspectors, although the agency's tight budget has to be taken into account.

Comprehensive nuclear test ban (CTB)

Since the entry into force of the NPT, many NNWSs have argued that the NWSs should agree on a comprehensive nuclear test ban. At the last review conference in 1990, many NNWSs called for serious negotiations on a CTB, and a few NNWSs, including Mexico, tried to link a test ban with the extension of the NPT. The United States strongly opposed any linkage and proposed a step-by-step approach.[6]

In October 1991 Russia announced a moratorium on testing, and in April 1992 the French government also proclaimed a moratorium. While the US administration has been reluctant to take steps towards a CTB, after the House of Representatives approved a one-year testing moratorium the Senate voted overwhelmingly on 3 August 1992 to suspend nuclear weapons testing for nine months, to limit the number of tests to fewer than five for the next three years, and to ban permanently all tests after 30 September 1996. Under such strong pressure from Congress and with a view towards winning the presidential election, on 2 October President Bush signed the bill passed by Congress. The bill contains a reservation that it remains effective so long as other NWSs refrain from conducting nuclear tests.

This presents a good opportunity for working towards a CTB, with Russia, France and the United States having suspended nuclear tests and the United Kingdom under a *de facto* moratorium. China, as a new member of the NPT, should cooperate with the other NWSs and negotiate a treaty on a CTB as soon as possible.

In the spring of 1993 the Clinton administration reportedly began to examine the possibility of resuming tests on very small nuclear weapons.[7] Such a step would be in opposition to current international public opinion which favours a ban. It will have an adverse effect on the NPT regime, for the following reasons. First, Russia and France might follow the US and resume testing. Second, US tests would give non-parties to the NPT an excuse to stay outside the treaty. Finally, NNWSs which are parties to the treaty would not agree to an indefinite or long-term extension.

In early July 1993 the United States, Russia, and France announced their decision to continue the moratorium on nuclear testing unless someone else tests first, and Britain supported the US decision. This is a big step towards a comprehensive nuclear test ban, and China should be persuaded to join the moratorium.

Security assurances to NNWSs

In order to strengthen the NPT regime and achieve increased participation by NNWSs, it is vital to take measures to strengthen the security of states that have voluntarily forsworn the nuclear option. Negative security assurances that prohibit the use or threatened use of nuclear weapons against NNWSs which are parties to the NPT have been sought by NNWSs since the treaty was negotiated. Although all five NWSs made declarations on negative security assurances in 1978 and 1982, at the first and second special sessions of the United Nations General Assembly on disarmament, their declarations are neither uniform nor legally binding. The NNWSs demand negative security guarantees that have no conditions and are legally binding.

A very favourable situation appears to be emerging for codifying negative security assurances. First, the intensity of the need for extended deterrence in the Cold War has disappeared with the improvement in East–West relations, the democratization of the former East European states, and the dissolution of the Warsaw Treaty Organization. Second, the need to supplement inferiority in conventional arms with nuclear weapons has also disappeared with the conclusion of the CFE treaty. Third, tactical nuclear forces have been withdrawn through the unilateral decisions taken by Presidents Bush and Gorbachev in the autumn of 1991. Finally, it has become possible for the five NWSs to negotiate on this problem within the framework of the NPT because China and France have joined it. With Ukraine, Kazakhstan and Belarus asking for some form of security assurance, this measure would play an important role in strengthening their commitment to the NPT. In addition, with the dramatic changes in the international security environment since the late 1980s, it seems that the time has come for the NWSs to negotiate a treaty on no-first-use of nuclear weapons.[8]

Positive security assurance, which means giving assistance when an NNWS party to the treaty is under threat of nuclear attack or under actual nuclear attack, was given through Security Council Resolution 255 (1968) at the time of the adoption of the NPT. In 1968 the People's Republic of China, which was not then a member of the UN, was thought to be a potential aggressor. However, since China obtained a seat on the Security Council in 1971, its veto power has meant that the resolution has been regarded as ineffective, or as amounting only to reconfirmation of the right to collective self-defence.

With the increase in potential or undeclared nuclear weapons states, Egypt, which is afraid of nuclear attack by Israel, demands positive security assurances that go beyond the provisions of Security Council Resolution 255.

As we enter the post-Cold War era, and with the prospect that the United Nations will revive its role in maintaining international peace and security, it is advisable to arrange for effective action by the United Nations under Chapter VII of the Charter. However, such action must be under the strict control of the Security Council.

To recognize positive security assurances to an individual state is dangerous and not recommended because it may increase the possibility of nuclear weapons being used and it would also enhance the political value of threatening to acquire nuclear weapons.

Nuclear-Weapon-Free Zones (NWFZs)

The establishment of NWFZs is a useful complementary measure to the NPT.[9] There have been many proposals for establishing NWFZs in various regions of the world, but their realization has been hindered by the Cold War. We should now increase our efforts to establish NWFZs in many regions, because favourable conditions for their introduction have emerged. First, in Africa, where establishing an NWFZ was first proposed in 1960, now that South Africa, the most dangerous potential nuclear state, has acceded to the NPT, the creation of an NWFZ appears a real possibility.

Second, in southeast Asia, where the general situation has improved since the withdrawal of Soviet troops from Vietnam, the closing of the US bases in the Philippines, and the declaration of a cease-fire in the Cambodian civil war, ASEAN states and other countries could profitably discuss the possibility of creating an NWFZ.

Third, in the Korean peninsula, US nuclear weapons have been withdrawn, North Korea has accepted IAEA safeguards, and both Korean states have agreed to denuclearization. This has made the peninsula a *de facto* NWFZ, and it may be possible to make it a *de jure* NWFZ. The situation of North Korea, however, is not yet sufficiently clear. It might also be possible to extend the NWFZ to include Japan, which maintains a threefold non-nuclear policy.[10]

The initiative for the establishment of NWFZs has to come from the NNWSs in the region, but cooperation of the NWSs is indispensable for successful implementation.

Nuclear export control

Nuclear Suppliers Guidelines Part I

In 1975 the nations of the London Nuclear Suppliers Group adopted guidelines for the export of nuclear-related material or facilities in order to prevent the spread of nuclear weapons technology. They have prepared a list of items which may only be exported in accordance with IAEA safeguards, and have restricted the export of items that could lead to proliferation.

After the revelation of the Iraqi nuclear weapons development programme and the dismantling of most of the COCOM controls, it was generally accepted that a new initiative was necessary to deal with the new situations.[11] In March 1991 member countries of the Nuclear Suppliers Group held an informal meeting in The Hague where they set up a working group to deal with dual-use items. The London Nuclear Suppliers Guidelines have played a role in non-proliferation, but they have certain weaknesses. First some states like France, Germany and the former Soviet Union have had a tendency to export for commercial benefit to the detriment of nuclear non-proliferation policy. Second, the criteria for export licences were not uniform, and each country applied the guidelines without taking into account the real proliferation risk in the recipient country. Third, dual-use items were not included in the trigger list: Iraq imported many dual-use items for the development of its nuclear weapons programme. Finally, the group did not provide for implementation of the guidelines; for example, there were no measures that could be taken against those who breach the guidelines.

Nuclear Suppliers Guidelines Part II

At a meeting of 27 members of the Nuclear Suppliers Guidelines in Warsaw from 31 March to 3 April 1992, there was agreement on guidelines for transfers of nuclear-related dual-use equipment and material, and related technology.[12] As an Annex, the meeting set forth a list of 65 items that are subject to the guidelines. As a basic principle they agreed that suppliers should not authorize transfers of equipment, material or related technology identified in the Annex for use in an NNWS for a nuclear-explosive activity or for nuclear fuel-cycle activities that are not safeguarded, or in general where there is an unacceptable risk of diversion to such activities, or where the transfers are contrary to the objective of averting the proliferation of nuclear weapons.

Export is forbidden for use not only in a nuclear-explosive activity but also in unsafeguarded nuclear fuel-cycle activities, placing the cause of non-proliferation above the promotion of international cooperation in the use of nuclear energy. Moreover, not only NNWSs but also NWSs are subject to the guidelines when they are importers.

With regard to the establishment of export-licensing procedures, in authorizing the transfers, suppliers should exercise prudence in order to carry out the basic principle described above. They should take the following factors into account: whether the recipient state is a party to the NPT, whether the transfer is appropriate for the stated end-use, and whether the recipient's actions and policies are supportive of nuclear non-proliferation. The list includes 65 items, most of which are dual-use items. Some items that are directly connected with nuclear weapons are included in spite of the NWSs' reluctance.

Under a Memorandum of Understanding, structural arrangements were agreed upon by the 27 members such as a mutual exchange of knowledge and information, a meeting for consultation, and the establishment of a secretariat in the Japanese embassy in Vienna. They also agreed to provide notice to the other countries of licences that have been granted or refused; rejection notices would be binding on all parties for a period of three years. The London Suppliers Guidelines Part II, agreed by 27 states (a sizeable increase from the original seven adherents to the London Suppliers Guidelines Part I in the 1970s), invite all states to adhere to the guidelines. There are still some states that have the capability to export nuclear-related items but which remain outside the group and it is necessary to persuade them to join it.

The criticism[13] that the guidelines are discriminatory (because they were created by the industrial exporting countries only and apply to transfers to the developing importing countries) still has some validity since only a minority of countries participated in making the rules. However, their discriminatory character has been mitigated by the emphasis placed on whether a recipient state is a party to the NPT, or is supporting non-proliferation objectives in transferring the listed items, and by subjecting transfers to NWSs to the export guidelines, in certain cases.

Finally, we have to examine the effectiveness of the guidelines from two aspects. Until recently, the guidelines in Part I had not been strictly implemented because some states compromised non-proliferation objectives in order to gain economic benefits, as was shown by the fact that France, Germany, the United Kingdom, the Soviet Union and others refused to apply full-scope safeguards in exporting nuclear-related items. At the meeting in Warsaw in April 1992, there was agreement that the transfer of nuclear facilities, equipment, components, material and technology, as referred to in the export trigger list of the Guidelines for Nuclear Transfer, should not be authorized to an NNWS unless that state had concluded an agreement with the IAEA on the application of safeguards on all source material and special fissionable material in its current and future peaceful nuclear activities.

This bodes well for significant progress in the effective implementation of the guidelines in Part I. Although we can anticipate that the element of commercial competition will decrease in importance, we must carefully track each state's practice in exporting the listed items. In addition, those states that have the capacity to export but have not joined the group must be persuaded to agree to the guidelines.

Second, technical barriers to proliferation will be effective in the short term but not in the long term.[14] Technological progress is rapid and information may be leaked; or a threshold country may be able to develop nuclear weapons without outside help. A variety of other measures are needed to prevent nuclear proliferation in parallel with export controls.

Conclusion

Although, among other shortcomings, the present Nuclear Non-Proliferation Treaty regime has a tendency to be discriminatory, it should be extended indefinitely or over the long term because it is indispensable to maintaining and improving the international security order beyond 1995.

With the implementation of nuclear disarmament among the NWSs and the emergence of new dangers for nuclear proliferation, it is becoming necessary to strengthen the regime. There are many ways of doing so, and these can be divided into political measures and technical measures.

With regard to technical means, the IAEA safeguards system should be applied more universally and should be strengthened by extending the power of the IAEA. The London Nuclear Suppliers Guidelines should be accepted by more countries and strictly implemented by all participants.

With regard to political means, nuclear disarmament measures that have been agreed should be implemented as soon as possible. The quantitative nuclear arms race should be stopped by a production cut-off of nuclear material for weapons purposes. And the qualitative nuclear arms race should be halted through a comprehensive nuclear weapon test ban. Negative security assurances should be provided through legally binding undertakings by all NWSs. The beginning of the post-Cold War era provides an opportunity for all these measures to be implemented.

Notes

1 For a general background to this problem, see: Lewis A. Dunn, 'Containing Nuclear Proliferation', *Adelphi Papers* 263, Winter 1991; Peter Clausen, 'Nuclear Proliferation in the 1980s and 1990s', in Michael T. Klare and Daniel C. Thomas, eds., *World Security: Trends and Challenges at Century's End* (St Martin's Press, 1991), pp. 144–69; John M. Deutsch, 'The New Nuclear Threat', *Foreign Affairs*, Vol. 71, No. 4, Fall 1992, pp. 120–34; Leonard S. Spector, 'Repentant Nuclear Proliferants', *Foreign Policy*, No. 88, Fall 1992, pp. 21–37; Donald G. Boudreau, 'On Advancing Non-Proliferation', *Strategic Review*, Vol. XIX, No. 3, Summer 1991, pp. 61–7; Joseph F. Pilat and Robert E. Pendley, eds., *Beyond 1995: The Future of the NPT Regime* (New York: Premium Press 1990); Harald Muller, 'The Nuclear Non-Proliferation Regime beyond the Persian Gulf War and the Dissolution of the Soviet Union', *SIPRI Yearbook 1992: World Armaments and Disarmament* (Oxford: Oxford Univ. Press, 1992), pp. 93–106; John Simpson and Darryl Howlett, 'Nuclear Non-Proliferation: The Way Forward', *Survival*, Vol. 33, No. 6, November/December 1991, pp. 483–99; Josef Goldblat, 'The Non-Proliferation Treaty: How to Remove the Residual Threats', *UNIDIR*, Research Paper No. 13, 1992.

2 See also George Bunn, Charles N. Van Doren and David Fischer, *Options and Opportunities: The NPT Extension Conference of 1995* (Southampton: PNN Study No. 2, 1991).

3 Thomas Graham Jr, 'The Duration of the Nuclear Non-Proliferation Treaty: Sudden Death or Long Lease on Life?', *Virginia Journal of International Law*, Vol. 29, No. 3, Spring 1989, p. 661.

4 See Carl Kaysen, Robert S. McNamara and George W. Rathjens, 'Nuclear

Weapons after the Cold War', *Foreign Affairs*, Vol. 70, No. 4, Fall 1991, p. 109.

5 McGeorge Bundy, William J. Crow Jr., and Sidney Drell, 'Reducing Nuclear Danger', *Foreign Affairs*, Vol. 72, No. 2, Spring 1993, p. 143.

6 Darryl Howlett and John Simpson, 'The NPT and the CTB: Linkages, Options and Opportunities', *Arms Control*, Vol. 13, No. 1, April 1992, pp. 85–107.

7 George Perkovich, 'Proliferation by Example', *Washington Post*, 5 May 1993.

8 Jack Mendelsohn, 'Arms Control after the Summit', *Arms Control Today*, May 1993, p. 14.

9 Jon Brook Wolfsthal, 'Nuclear-Weapon-Free Zones: Coming of Age?', *Arms Control Today*, March 1993, pp. 3–9.

10 The threefold non-nuclear policy means that Japan should neither possess, manufacture, nor permit the introduction of nuclear weapons on its territory.

11 Paul J. Leventhal, 'Plugging the Leaks in Nuclear Export Control: Why Bother?', *Orbis*, Vol. 36, No. 2, Spring 1992, pp. 167–80.

12 INFCIRC/254/Rev.1/Part 2, July 1992.

13 K. Subramanyam, 'Export Controls and the North–South Controversy', *Washington Quarterly*, Vol. 16, No. 2, Spring 1993, p. 135.

14 Jean-François Rioux, 'Supply-Side Controls on Nuclear Proliferation', *Orbis*, Vol. 36, No. 2, Spring 1992, p. 164.

CHAPTER 5

CONTROLLING MISSILES, AND CHEMICAL AND BIOLOGICAL WEAPONS

PETER VAN HAM

Introduction

For obvious reasons, Western countries have paid most attention to the problems of the proliferation of *nuclear* weapons. Nuclear weapons are exceptionally destructive, and are therefore generally considered as forming a class of their own. Other categories of weapons of mass destruction, such as chemical and biological weapons,[1] are of a markedly different nature. In contrast to nuclear weapons, they produce no collateral damage, and they require materials that are widely available from peaceful facilities and are therefore relatively easy to manufacture. Chemical weapons are furthermore distinctive because they have been used in combat situations since 1945. Because of their destructive power and their indiscriminate effects, chemical and biological weapons have evoked deeply felt aversion, both within the military and with the general public.

This chapter will examine the threat posed by chemical and biological weapons, as well as the means to deliver non-conventional weapons by ballistic missiles. It will analyse the development of the relevant control regimes, and point out their major characteristics in terms of membership and operating procedures. It will further analyse the defects and strengths of these regimes, and it will suggest some possibilities for reinforcing them. Finally, it will assess the usefulness and feasibility of integrating these mechanisms in an all-embracing control regime.

Chemical weapons

The threat of chemical weapons is an old one. The Greeks, for example, first used sulphur mixtures against their enemies as early as 431 BC. Early modest attempts to outlaw the use of chemical weapons include the 1874

Brussels Declaration on Laws and Customs of War, prohibiting the use of poison and poisoned weapons. During the First World War, however, chemical weapons caused some 1.3 million casualties, of which more than 91,000 were fatal. These experiences resulted in the 1925 Geneva Protocol prohibiting the use of chemical weapons. The Geneva Protocol did not, however, forbid the stockpiling of such weapons, nor did it restrict the research, development, production, testing, and transfer of these weapons. The Geneva Protocol's immediate impact on state behaviour proved limited, as Italian troops used chemical weapons during their attack on Abyssinia in the 1930s. Experience during the Second World War was more encouraging: although both sides had chemical weapons, there was tacit agreement not to use them. However, owing to the Geneva Protocol's limited authority and muscle, the production and stockpiling of chemical weapons has continued over the years.

Apart from the occasions referred to above, chemical weapons were hardly used until the 1980s. This may have been because of their debatable military usefulness. For long they were perceived as having only limited utility in warfare. Their dependence upon the prevailing meteorology (especially wind) limited their accuracy and general utility, potentially posing serious risks for the aggressor's troops. However, chemical warfare agents which act within a few minutes have a potential use in near-contact battles. Protective measures against chemical warfare can be taken, but when used against unprotected personnel, chemical weapons can be extremely hazardous.

Development of control regimes

In the 1980s, the Western world was alarmed by reports that countries like Egypt, Iraq, Libya, and Syria were constructing massive production facilities for chemical weapons. Countries like North Korea and Taiwan were also suspected of having chemical weapons stocks. Previously, the West had only shown concern about Soviet chemical (and biological) weapons programmes and their threat to NATO military forces. But after the Cold War, the proliferation of chemical (and biological) weapons among developing countries quickly became an issue of major political concern. US and German intelligence sources have reported that Libya is building its second chemical weapons plant, and that as much as 100 tonnes of chemical weapons are already stored at Rabta, its first chemical weapons factory.[2] The former CIA Director Robert Gates also argued on 15 December 1992 that 'Iran has an active chemical weapons programme ... It has produced at least several hundred tons of blister, choking and blood agents – and possibly as much as 2,000 tons – at steadily increasing rate since 1984.'[3]

It was the actual use of chemical weapons which provided the alarm for the international community to strengthen the Geneva Protocol, and to develop a multilateral approach to verifying and enforcing their elimination. The use of chemical weapons in the Iran–Iraq war of the mid-1980s (which resulted in an estimated 10,000 casualties) placed these weapons high on the international agenda and has been the main catalyst for action. It soon became clear that several Western governments and Western chemical companies had played a major role in providing key components and commodities for Iraq's chemical weapons programme. Western export controls to limit the proliferation of chemical weapons had proved insufficient, and in 1985 – under the auspices of Australia – fifteen chemical supplier countries decided to implement stricter export controls on key precursor chemicals, and to harmonize their export controls on commodities related to chemical weapons and development mechanisms. The main goal was to restrict the spread of chemical weapons by making it more difficult and costly for (potential) proliferators to obtain the chemical commodities, technology and services which are required for weapons production.[4] The official purpose of the so-called Australia Group is to explore the scope for making the export control measures of the participating countries more effective, for example by the exchange of information, the harmonization of measures already taken, and, when necessary, consideration of the introduction of additional national measures. The Australia Group (now with 24 members), comprises Argentina, Australia, Austria, Canada, Finland, Hungary, Japan, New Zealand, Norway, Sweden, Switzerland, the United States and all twelve European Community countries (including the EC Commission). Meetings are held on a biannual basis (in the Australian embassy in Paris). A number of countries in Eastern Europe have now imposed similar controls, and in response to their requests for closer contact with the Australia Group, several seminars on the problems of chemical (and biological) proliferation have been organized. Most of these countries have introduced wide-ranging national export restrictions on materials, technology and services which are relevant to the production of chemical (and biological) weapons. Like all other supply-side regimes, a number of important producers are not participating in the Australia

Group: for instance China, India, Mexico, South Africa and South Korea. China, notably, is reported to have sold chemical precursors to problem countries (e.g. Libya).[5] Beijing, however, did sign the CWC in early 1993 (see below).

Within the Australia Group, formal licences are now required to export a 'core list' of eight particular chemicals, and a so-called 'warning list' of another 50 (mainly dual-use) chemicals has been compiled; chemical industries have been requested voluntarily to alert their national governments of foreign interest in these chemicals. Although no chemical or biological weapons were used during the Gulf war, the threat of their use played an important role. What is more, almost daily television reports of the population of Tel Aviv, preparing with gas masks for Iraqi Scud missiles perhaps carrying chemical warheads, focused public attention on the threat posed by these weapons. In May 1991, after the Gulf war had come to an end, the participating countries of the Australia Group gathered to strengthen their export control regime. All members pledged to require export licences for all 50 chemical products on the warning list (except for exports to other Australia Group members).[6] The Australia Group also agreed upon controls on a list of chemical weapons-processing equipment. Earlier, in mid-1990, the Australia Group had already expanded its scope to cover biological weapons, issuing warning guidelines for equipment, goods and technological data for the manufacture of biological agents. During its December 1992 plenary session, the Group adopted control lists of human and animal pathogens and biological equipment. It also developed an 'industry awareness' list of plant pathogens.

The gassing of Kurdish villagers including women and children by Iraqi troops at Halabja in March 1988 (which received much media attention in the West) has played a major role in intensifying existing efforts to address the proliferation of chemical weapons on a global basis. French President Mitterrand put forward a proposal in a speech to the United Nations General Assembly, in September 1988, for an embargo on trade in chemical commodities and technology with Iraq, to prevent the use of chemical weapons.[7] Later that year, US President Reagan called for an international conference that would strengthen the Geneva Protocol. The most appropriate forum to address this issue was an Ad Hoc Committee of the Conference on Disarmament (CD) in Geneva, where international diplomats were already negotiating over the text of a global Chemical Weapons Convention (CWC). The ultimate goal of the CWC was to ban the development, production, stockpiling and use of these weapons. It would also design mechanisms for verification and enforcement. For many years, the East–West conflict and North–South distrust had made such an agreement impossible. But the thawing of the Cold War has opened the way for further cooperation, and the realization of developing countries that they would be the most likely victims of chemical weapons significantly increased the urgency of completing the CWC.

The Chemical Weapons Convention

After many years of negotiations, a final CWC text was agreed in early September 1992. The treaty was signed by 142 states in Paris in mid-January 1993, and now awaits worldwide ratification.[8] Among the signatory states are many former Soviet republics and, perhaps somewhat surprisingly, states like China, Iran, Vietnam, Israel, South Korea, India and Pakistan, which were suspected of possessing chemical weapons capabilities.[9] Each state party to the treaty is legally obliged never in any circumstances to:

- develop, produce, otherwise acquire, stockpile or retain chemical weapons, or transfer, directly or indirectly, chemical weapons to anyone;
- use chemical weapons, or engage in any military preparations for doing so; or
- assist, encourage or induce, in any way, anyone to engage in any activity prohibited by the treaty.

The treaty will enter into force in January 1995, or 180 days after the 65th state has ratified it, whichever is later. Parties will have until 2005 to destroy the chemical weapons they may now have (only the US, Russia and Iraq are acknowledged chemical weapons states). An Organization for the Prohibition of Chemical Weapons (OPCW) is to be headquartered in The Hague, and will have somewhat similar tasks to the IAEA (although it has greater authority for monitoring the CWC's functioning than the IAEA with respect to the NPT).[10] The OPCW is made up of three organizations. First is the Conference of States Parties, which includes all the parties and is the 'principal organ' of the OPCW. It will meet annually and also in special sessions when there is need to do so. The conference has responsibility for taking 'the necessary measures to ensure compliance' and for redressing violations, and it has the power to suspend a violator's rights and privileges. Second comes the Executive Council, which will have a rotating membership of 41 states chosen by the conference for two-year terms, distributed regionally. The Executive Council is the 'executive organ' of the OPCW, responsible for overseeing the implementation and operation of the treaty. Third is the Technical Secretariat, which includes an Inspectorate as well as a Scientific Advisory Board of independent experts. The secretariat is responsible for the OPCW's day-to-day work, carrying out all the details of the implementation and verification of the CWC, including the sensitive task of negotiating arrangements for the conduct of inspections in each state. During the 1993–5 interlude, a Preparatory Commission (PrepCom) will operate,[11] preparing for the establishment of the OPCW, the recruitment and training of staff (about 300 to 400 specialists plus support personnel), and the purchase, standardization and testing of equipment.[12]

The level of control over individual chemicals relates to the level of risk that the chemical may be used for military purposes. Three so-called 'Schedules' have been identified for different groups of chemicals:

- *Schedule 1 chemicals* have toxicities that *would* enable them to be used as chemical weapons, or they may be immediate precursors to such chemicals; they pose a *high* risk to the object and purpose of the CWC. Schedule 1 chemicals have little or no use for purposes not prohibited under the CWC;
- *Schedule 2 chemicals* have toxicities that *could* enable them to be used as chemical weapons, or are immediate precursors to such a chemical or a precursor to a Schedule 1 chemical; they pose a *significant* risk to the object and purpose of the CWC. Schedule 2 chemicals are not produced in large commercial quantities for purposes not prohibited under the CWC;
- *Schedule 3 chemicals* have toxicities that *might* enable them to be used as chemical weapons; hence they pose a risk to the object and purpose of the CWC. Schedule 3 chemicals may be produced in large commercial quantities for purposes not prohibited under the CWC.[13]

Twelve families of chemicals have been identified as prime precursors for chemical weapons (Schedule 1), and manufacturing these chemicals is banned (except for very small quantities under close surveillance of the

OPCW Technical Secretariat). The international transfer of Schedule 2 chemicals is prohibited except between CWC states. Other chemicals that are frequently used in commercial products (Schedule 3) may be transferred freely. The CWC stipulates that it does not prohibit the production of chemicals for industrial, agricultural, research, medical, pharmaceutical or other peaceful purposes; for purposes related to protection against chemical weapons; for military purposes not related to the use of chemical weapons; and for law enforcement (including domestic riot control).

Clearly there is a rather thin line between these legitimate purposes and those prohibited by the CWC. Many chemical weapons are made of different commodity chemicals which are not all controlled by the Australia Group. A country determined to acquire chemical weapons could therefore import these chemicals separately, claiming that they were for legitimate peaceful purposes, and then combine them into weapons. Determined proliferators can also use so-called 'front' companies and false documents to import the required chemicals to produce the weapons. Literature on producing chemicals for weapons is widely available. The dangers of hidden stockpiles will remain, as well as the possibility of swift conversion of factories for peaceful production into chemical weapons production. Furthermore, it will be very difficult to monitor all the plants that produce legal pesticides, fertilizers, pharmaceuticals and similar products.

What makes the CWC exceptional and very ambitious is its system for international verification and compliance. Two verification systems will be set up. First, routine inspections will be carried out by the Technical Secretariat to verify information provided by participating states on particular issues specified in the treaty. Those data will be verified by regular on-site inspections. The number of routine inspections is, however, limited, since no single plant (involving Schedule 1 chemicals) is required to be inspected more than twice annually, and no state has to host more than 20 routine inspections per year (with a minimum of three inspections per year). The other system includes so-called 'challenge inspections', whereby the challenged party must grant 'the greatest possible degree of access.'[14] A maximum of 108 hours may pass after the arrival of the challenge inspection team before the host state has to give the inspectors internal access to the suspect site. The inspectors will work on a so-called 'managed access' basis, which signifies that host states have the right to take measures to protect their national and/or commercial interests. This certainly limits the intrusiveness of the inspections, and may therefore in certain circumstances increase the possibility of cheating. As one observer argued: 'Because access can be denied on plausible trade or national security grounds, the information obtained by inspection teams may be suggestive rather than conclusive.'[15] One could therefore argue that the routine inspections are not more than confidence-building measures, whereas the challenge inspections go further, and provide a substantive mechanism for deterring cheating.

If the routine or challenge inspections raise doubts as to whether a party is complying with CWC obligations, a deadline may be set for the suspect party to provide assurances. If this does not happen in a satisfactory manner, sanctions may be imposed. Particularly grave offence against the CWC, or obstruction during inspections, may be referred to the UN General Assembly and/or the UN Security Council. The OPCW can be requested by parties to investigate suspected sites, and make on-site challenge inspections. The Executive Council has the power to block a challenge inspection by a three-quarters majority, if it determines that such an inspection is 'frivolous, abusive, or clearly beyond the scope' of the CWC.

Although the CWC is a major breakthrough, there are still several obstacles to be overcome. For example, the ratification process is not assured in all signatory countries. Russia may postpone ratification of the CWC until international financial assistance for the Russian chemical weapons destruction programme is guaranteed (see below). Another problem is the fact that several countries remain outside the CWC, including Egypt, Taiwan, North Korea, Libya and Syria. A minority of Arab countries including Egypt and Syria have declared that they will only sign the CWC when Israel agrees to join the NPT as a non-nuclear weapons state and eliminates its nuclear arsenal.[16] The (former) Egyptian Foreign Minister Abdel Meguid, has argued: 'We consider that progress in the field of the prohibition of chemical weapons ... is linked to the realization of a parallel prohibition in the field of nuclear weapons [by Israel].'[17] Some Middle Eastern countries are not convinced that giving up their chemical weapons without compensation will benefit their national security situation, especially since Israel's nuclear weapons will then go unchallenged. Considering the improbability of Jerusalem joining the NPT, it is difficult to see how certain Arab countries can be persuaded that signing the CWC is worthwhile.

The CWC does include a combination of sticks and carrots. States facing chemical threats or attack are entitled to receive assistance (including defensive equipment such as sensors, protective clothing, decontamination equipment, antidotes and advice on chemical defensive measures). CWC member states are also encouraged (by Article XI) to 'participate in the fullest possible exchange of chemicals, equipment, and scientific and technical information ... for purposes not prohibited'. The CWC's sticks include restricting chemical trade with non-signatories, although it is unlikely that the sanctions for non-compliance will be compelling enough to change these countries' intentions. The CWC stipulates that countries that refuse to commit themselves to the treaty will face trade restrictions on dual-use chemicals. Countries that do not join are banned from trade in Schedule 1 chemicals. They will also be required to provide end-user certificates guaranteeing that controlled chemicals will not be used for military purposes. Three years after the CWC enters into force, Schedule 2 chemicals may no longer be exported to non-signatory countries. Five years after the treaty enters into force, the Conference of States Parties must consider whether more stringent controls on the transfer of Schedule 3 chemicals are required.

Defects and strengths

The CWC's verification procedures are among the broadest and most intrusive of multilateral arms control agreements. Although the CWC has not had time to prove its utility and force in practice, it is possible to point out several likely flaws and merits.

The very nature of many chemicals makes them dual-use (with the notable exception of those families included in Schedule 1). No CWC will therefore ever be 100 per cent verifiable. Ways of cheating are numerous. We have already indicated how determined proliferators may import certain chemicals for peaceful purposes, and then divert them into chemical weapons. Most diversions of common chemicals which are legitimately produced in very large quantities cannot be detected (or detection is at least extremely difficult). Practical problems with examining possible diversions also abound, since accounting errors may be made, chemicals may evaporate, be spilt or be mismeasured, and records may be falsified. There is furthermore an inherent tension between verification and commercial interests which might in practice limit the CWC's intrusiveness. Verification might jeopardize legitimate commercial activities in the chemical industry, where some level of secrecy is required to maintain a competitive edge.

The almost global coverage of the CWC also implies that a very large number of facilities will be subject to inspection: this makes it unlikely that all sites of potential interest will be inspected regularly enough to nullify the chance of cheating. Moreover, inspection teams will also want to know where to look in the first place. Detailed and reliable intelligence will be essential for the OPCW's operation, but it is not at all clear how it will obtain the information to enable it to decide which facilities warrant inspection. A computer database will be established in The Hague comprising information provided by the member states. The OPCW will, however, not have an independent information-gathering network.[18] This might prove a significant lack, since apart from controlling declared chemical production facilities, undeclared stock and covert production must also be discovered. Problems of the UNSCOM in Iraq, and the IAEA in South Africa and North Korea, with discovering nuclear and chemical weapons facilities and determining the quantities of sensitive materials, have already demonstrated how difficult it is to locate suspect sites and quantify stockpiles. What is more, chemical inspections will be very costly, at approximately US$1 billion per year for the CWC (whereas the IAEA 'only' has a US$160 million annual budget),[19] and will require remarkable political commitment. Given their other responsibilities, it is unclear whether countries like the United States and Russia will be prepared to provide the CWC and the OPCW with sufficient financial resources. The destruction of the US stock of 30,000 tonnes of chemical agent is estimated to cost some US$7.9 billion. It is also probable that the West will end up paying most of the cost of destroying Russia's 40,000-tonne stockpile (estimated at US$10 billion).[20] Chemical weapons will have to be destroyed in an environmentally safe way, and this will be closely monitored by environmental pressure groups like Greenpeace. This will add to the cost of implementing the CWC, and will make the realization of the treaty's goals far from assured.

The merit of the CWC is that it delegitimizes the production, stockpiling and use of chemical weapons on all occasions. The CWC provides the OPCW with extensive legal authority, which it should be prepared to use with determination. As Amy Smithson correctly argued: 'The OPCW will lose credibility if it follows the IAEA's footsteps; that is, starting out with inspections limited in scope and remaining static over time ... To succeed in the verification business, the OPCW must have a strong and steady start.'[21] Strong backing by the United Nations and its Security Council for the CWC and the OPCW's inspections will therefore be required, especially since the PrepCom will be pressurized by states that are apprehensive about strict on-site inspections and would like it to slacken the verification regime.

There can be little doubt that the CWC is not – and never will be – 100 per cent watertight: determined proliferators will always be able to manufacture chemical weapons. However, it can be argued that the CWC provides significant verification procedures which deter cheating, and that its mixture of verification, confidence-building measures, assurances and sanctions may go a long way in encouraging states to abandon this category of weapons. The CWC will most certainly increase the costs for states to manufacture and stockpile chemical weapons, and it is also likely to make the covert development of large-scale, militarily significant chemical weapons capabilities less feasible. It is furthermore important to stress the fact that the CWC will not be a discriminatory regime, but that it will make a serious attempt to ban all chemical weapons. The CWC lacks the clear-cut North–South cleavage which makes other non-proliferation mechanisms (like the Australia Group and the MTCR, discussed below) so unacceptable to many developing countries.[22] From this perspective, experiences with the CWC may provide some useful lessons for the NPT and the BTWC (see below), and may reinforce the demand of developing countries that nuclear weapons countries should abolish such weapons.

Biological weapons

It is clear that biological weapons are closely related to chemical weapons, since both (1) harm human beings rather than inflict damage on property; (2) are effective over large areas; (3) have an indiscriminate impact; and (4) are relatively cheap to produce. Biological warfare agents work more slowly than chemical weapons because of the need for the micro-organisms to replicate themselves inside the victim's body. They therefore might have different uses in battle situations, for instance against rear areas and in static situations.

Biological weapons have never been openly used in military conflicts, but their possible strategic and tactical uses have been researched in depth by military planners. As with chemical weapons, their military utility is dubious. Many regard biological weapons as slow and unpredictable. The supposed practical problems with using this kind of weapon in combat (like the influence of weather conditions) make them less of a strategic asset. Developments in genetic engineering and biogenetic technologies, however, have opened up new possibilities for biological weapons, making them potentially even more destructive than naturally developing organisms. But it is still unlikely that this category of weapons has a rational military mission, outside the limited field of so-called 'covert operations'. Since biological weapons have never been used, considerably less attention has been paid to establishing mechanisms for their control.

Development of control regimes

The Convention on the Prohibition of the Development, Production and Stockpiling of Bacteriological (Biological) and Toxin weapons and on their Destruction (the BTWC), was opened for signature in 1972, and came into force in 1975. The United States, the Soviet Union and most other major countries have ratified the BTWC (as of January 1992, 118 states were party to it), thus making possession of biological weapons illegal, 'whatever their origin or method of production, of types and in quantities that have no justification for prophylactic, protective, or other peaceful purposes'. The taboo on this kind of weapons has been immense, and this has probably contributed to their relatively limited proliferation. The BTWC does, however, permit the development and production of biological weapons for defensive purposes, and it does not prohibit the research and manufacture of biological weapons equipment, the evolution of biological weapons warfighting doctrines, and protective training by armed troops.

Every five years, the BTWC has a Review Conference. During the 1986 (Second) Review Conference, four confidence-building measures (CBMs) were agreed:

- declaration of all BL4 (maximum containment) laboratories and of all defence BL3 and BL4 laboratories;
- declaration of unusual outbreaks of disease;
- encouragement of publication of results; and
- encouragement of international contacts and conferences.

The final declaration of the 1991 Review Conference added other CBMs (which include the declaration of past activities in offensive and defensive biological research and development programmes, and the declaration of vaccine production facilities), and reaffirmed the crucial importance of all elements of the BTWC. It is also proposed to hold a Fourth Review Conference not later than 1996. An Ad Hoc governmental Group of Experts was mandated to identify and examine potential verification measures from a scientific and technical standpoint. This Ad Hoc Group will identify measures for establishing whether a state party is developing, producing, stockpiling, acquiring or retaining biological weapons of types and in quantities that have no justification for prophylactic, protective or peaceful purposes, as well as the equipment (or the means of delivery) designed to use such agents for hostile purposes or in armed conflicts. The first Ad Hoc Group meeting (the so-called VEREX meeting) was held in March 1992, and further meetings took place in 1993.[23] The Ad Hoc Group is preparing a report which will include proposed verification measures to be discussed during a conference of the BTWC state parties.

Additional steps to control these weapons had been taken in December 1991, when the Australia Group decided to extend its scope to cover biological weapons (see above). Owing to the sometimes unclear formulation of the BTWC articles, there have been problems with the implementation of the regulations in the domestic legislation of the states party to the BTWC. What is more, the problems of verification are also considerable. Since most biological research facilities and even modestly developed pharmaceutical industries can easily produce materials for biological or toxin weapons, verification is almost impossible. This, of course, considerably limits the impact which any biological non-proliferation regime might have. It will be difficult to construct a regime which will provide a solid global framework of cooperation, halting the proliferation of biological weapons.[24] As the US ambassador to the BTWC Review Conference of 1991 argued somewhat pessimistically: 'the Convention is not effectively verifiable and we do not know any way to make it so.'[25]

Among the mechanisms for promoting cooperation within the BTWC are CBMs whereby information is exchanged and consultative meetings are held. The proposals of the Ad Hoc Group will be most influential for the biological weapons control regime of the future, setting important steps towards a global framework. It is to be hoped that the initiatives of the Ad Hoc Group will stimulate the public debate on how best to address the threat of biological weapons. It would be somewhat ironic to wait for biological weapons to be actually used in combat (as was the case with chemical weapons) before taking further and more decisive steps towards establishing a multilateral control regime. In the meantime, it has been argued that the OPCW should undertake the task of verifying the BTWC's operations. It should, however, be noted that the knowledge required for verifying chemical and biological weapons disarmament is very dissimilar. Although there is much to be said for strengthening the operational links between the two regimes, combining them might be rather premature (see below). It would also add considerably to the costs of the OPCW.

It may well be that the experiences of the CWC – and especially the practice of verification – will provide some useful lessons for constructing a regime aimed at limiting the spread of biological weapons. When the CWC is ratified, and if the OPCW operates satisfactorily, pressure to establish a similar regime for biological weapons is likely to increase. To be sure, stepping up the preparations for creating a CWC-style verification regime which could be adopted during the 1996 Review Conference should be a priority. Since the UN Special Commission (UNSCOM) established under UN Security Council Resolution 687 has also been charged with the elimination of Iraq's biological weapons, something might be learned from that experience too: UNSCOM has worked out a compliance-monitoring regime which could provide a starting point for a future verification regime for the BTWC.[26]

Ballistic missiles

The 1970s and 1980s were decades of dangerous proliferation of missile technologies due to the export activities of major states.[27] The main hazard of the spread of missile technology has been its so-called 'multiplier effect', in the sense that any country with nuclear and/or chemical and biological weapons capabilities will dramatically increase its military strength after acquiring the ability to deliver these weapons over a great distance.

In response to this mounting threat of ballistic missiles, an export control regime has been established under the name of the Missile Technology Control Regime (MTCR). The MTCR was originally set up in mid-April 1987, when the Group of Seven (G7) industrialized countries (Canada, France, Italy, Japan, the United States, the United Kingdom and West Germany) announced that they would restrict exports that could advance the proliferation of nuclear-capable missile technologies (with a range of 300 kilometres or more, and a payload of 500 kg or more). On the initiative of the United States, secret negotiations on such a control system had already started in November 1982, and informal implementation of several elements of a common export control policy had begun in 1985.[28] Since then, the MTCR has been joined by seventeen other states (Argentina, Australia, Austria, Belgium, Denmark, Finland, Greece, Hungary, Ireland, Luxembourg, the Netherlands, New Zealand, Norway, Portugal, Spain, Sweden and Switzerland).[29] In May 1990 the USSR announced that it would observe the spirit as well as the guidelines of the MTCR, and that it was willing to take measures to restrict the proliferation of missile technology. Moscow's step was of major importance, since the USSR was by far the greatest exporter of ballistic missiles. The Russian successor government of the USSR has agreed to follow a similar policy in the years ahead. However, the commitment of major missile producers like China and North Korea is still in considerable doubt.

It is important that, contrary to the CWC and the NPT, the MTCR is based neither upon a treaty nor upon an official agreement. It also lacks an international executive body to monitor the exports of missile-related technologies on a permanent basis. It has no verification provisions and no mechanisms to enforce its guidelines, and cannot apply sanctions to signatory or non-signatory countries which violate the guidelines. The MTCR thus has little binding force. However, many proscribed missile technologies are also on the COCOM list, and the MTCR relies on the national licensing procedures of its members to control the transfer of missile technology; it also uses COCOM categories for the classification of its items. The MTCR is therefore not entirely without teeth.

The MTCR currently has two basic thrusts: (1) a 'strong presumption' to deny the export of complete missile systems as well as major subsystems (which, taken together, are called Category I systems); and (2) less rigid restrictions on the export of dual-use missile technology and components (Category II systems), which are judged on a case-by-case basis. By focusing its controls on the Category I systems, the MTCR has tried to curb the export of only a limited list of crucial components for missile manufacturing in the hope that this will be sufficient to halt proliferation. On 7 January 1993, the MTCR members declared that the guidelines would be extended to missiles capable of delivering biological and chemical weapons (going beyond its previous scope of nuclear weapons). This was mainly in response to the experiences of the 1991 Gulf war, which had heightened concern that missiles might carry chemical or biological warheads weighing less than 500 kg, the estimated weight of a rudimentary nuclear warhead. A further MTCR plenary meeting was held in Switzerland from 30 November to 2 December 1993.

It is generally realized that neither the MTCR nor other non-proliferation systems in place today can guarantee that a country with sufficient financial resources and zeal will not acquire ballistic missiles. The MTCR has nevertheless had some limited success in hindering or slowing down several ballistic missile programmes, gaining time for the international community to put pressure on several threshold countries. As a result of pressure from the United States, cooperation between Argentina, Egypt and Iraq to construct the Condor II collapsed when the leaders of the first two states announced the termination of the programme. The MTCR has also hindered missile proliferation by making it more costly. In 1988, the director of the Brazilian Space Activities Institute (IAE) complained that the MTCR's export restrictions had increased Brazil's costs by US$8.4 million, and had delayed the first launch of its rocket by two years.[30] This indicates that the MTCR has been effective in delaying a missile programme and in raising the costs of the acquisition of missile-related items. It is, however, like all other supply-side non-proliferation regimes, unable to halt a politically determined threshold state from obtaining the required components for its missile programme.

Defects and strengths

Three major weaknesses are ingrained in the MTCR. First of all, the MTCR has a limited membership: China and North Korea, notably, do not show signs of restricting their exports of missile systems and components, although Beijing has agreed to abide by the MTCR's guidelines. There are, furthermore, several new supplier states which are developing space launch technology primarily for commercial, peaceful purposes, but whose lack of rigid export controls might contribute to the spread of ballistic missile capabilities. Second, there is no psychological or ethical taboo on the export or development of ballistic missiles comparable to that on nuclear, chemical and biological weapons: no moral stigma is attached to the export of missile technology. This makes a strong regime difficult to maintain. The third problem is that the harmonization of the export restrictions is complex, and not all participating countries apply the MTCR guidelines in the same manner and with the same rigour. The MTCR's lack of an executive and coordinating body also makes it very difficult to interpret the export restrictions, which in practice leaves undue leeway for national discretion. This has led to difficulties because some countries allow the export of technologies to the commercial disadvantage of other members. Several countries continue to provide dual-use technologies to friendly states – be it for commercial or foreign policy reasons – without, however, sufficient checks on the end-use of their exports. Several MTCR members (especially France) have shown only limited willingness to restrict their exports of missile-related technology. In 1989 France decided to supply Brazil with liquid-fuel propulsion technology which could be used for ballistic missiles. Several West European countries have also been involved in exports assisting the Iraqi missile programme.[31]

The threat posed by ballistic missiles (and the warheads they deliver) will increase in the 1990s. The MTCR has a major role to play in managing the proliferation problems, but it has to be acknowledged that indigenous modernization of existing missile capabilities, as well as the spread of scientific and technical know-how and skills, will make this task very difficult. By focusing its control efforts on the crucial (Category I) components, the MTCR has made itself somewhat less unacceptable to developing countries. However, this leaves unchanged the fact that it will constrain the space programmes of those countries, it will deepen the North–South techno-divide, and it will be perceived by many Third World political leaders as impeding their economic development.

Options for addressing new proliferation challenges

It is evident that all approaches towards limiting the spread of non-conventional weapons and ballistic missiles have to be both multilateral and multi-pronged. International cooperation is crucial, as is the development of a blend of policies which tackles both the supply side and the demand side for these categories of weapons. One worthwhile approach towards limiting the spread of non-conventional weapons and ballistic missile technology and know-how has been the construction of science centres in the former Soviet Union. In early 1993, an International Science and Technology Center (ISTC) was established in Moscow, jointly funded by the United States, the European Community and Japan,[32] which now employs a number of Russian nuclear scientists. A similar centre is due to open shortly in Kiev. The main purpose of these centres – according to an official statement – is to 'minimize any incentives [for weapon scientists and engineers] to engage in activities that would result in proliferation of nuclear, biological and chemical weapons and missile delivery systems'.[33]

These ad hoc measures are essential for limiting the spread of non-conventional weapons. But one could also ask whether it would be useful and feasible to create a more consolidated non-proliferation structure, integrating some of the existing control mechanisms in an all-embracing control regime.[34] Recently, a study by three specialists in the field, Ashton Carter, William Perry and John Steinbrunner, argued that integration of the existing non-proliferation regimes could strengthen all regimes 'by unifying control bureaucracies and procedures for efficiency, sharing verification and transparency measures synergistically, and depoliticizing individual applications of the controls'.[35] The study maintains that these regimes already have much in common; for example, they involve similar legal, regulatory and administrative issues, are subject to similar political and commercial pressures, deal with the same arms traders, and the same security dilemmas animate the efforts of countries to acquire these weapons of mass destruction. Consolidating several regimes might apparently have substantial merit. Efficiency could be improved: for example, by enhancing the exchange of data among licensing officials, the costs of managing these regimes could be limited. Consolidation could also strengthen the efforts to limit the spread of non-conventional weapons by raising public moral concern, and by increasing awareness within firms.

The CWC and the BTWC are the regimes most likely to be combined at the treaty level. Both regimes ban the production, stockpiling and use of their respective weapons category for all state parties. The NPT differentiates the five acknowledged nuclear weapons states (the United States, Russia, France, Great Britain and China) from other member states. Including the NPT in such a 'super-regime' would be difficult, particularly since this would jeopardize the extension of the NPT beyond 1995 (see Chapter 4). Including supply-side export control mechanisms like the Australia Group, the NSG, COCOM or the MTCR in such a global, comprehensive treaty also seems unfeasible. In general, these export control frameworks are little more than 'gentlemen's agreements', which have been established to address a concrete proliferation threat, and which all have different membership rolls (although there is considerable overlap between them). What is more, target states would generally be located in the Third World, and many developing countries would regard such a super-regime as an organization aimed at excluding them from modern weaponry and high technology.

On verification, it may be possible to expand the ties between the IAEA and the OPCW. Other regimes lack such verification provisions and organizations. A single agency verifying compliance with nuclear, chemical and biological non-proliferation treaties may be feasible but for various practical and formal reasons is not on the cards for the foreseeable future. It might be imagined that teams of inspectors could carry out several tasks during one mission, but the detailed knowledge and practical details for the various regimes differ considerably, and inspections are therefore not easy to combine. The weapons technologies, production requirements, deployment and the consequences for the civilian industry of all these regimes are unique to each of them.

Some cooperation at the export control level has already materialized, with the Australia Group's decision to expand its controls to biological weapons materials and production equipment. There are several possibilities for creating a single agency which would be responsible for managing the multilateral export control aspects, but since this would require a single list of members, a unified list of controlled items and regime-wide licensing rules, many problems would have to be overcome to reach this goal. Since many regimes are targeted against countries that are members of other export control regimes, developing a single membership list seems impossible. It would also result in complaints from Third World countries, and rightly so.

In conclusion, it seems impractical and undesirable to combine several non-proliferation regimes into one comprehensive regime. However, more intensive coordination among these regimes may be sensible, and exchanging experiences of verification and enforcement will undoubtedly be helpful in generating more effective and efficient non-proliferation measures.

Notes

1 Chemical warfare agents are generally defined as chemical substances (whether gaseous, liquid or solid) which might be employed because of their direct toxic effect. Biological warfare agents are defined as living organisms which are intended to cause disease or death, and which depend for their effect on their ability to multiply in the organism attacked. Toxins are chemical substances produced by biological systems, but are not themselves living. See Brad Roberts, *Chemical Disarmament and International Security* (London: Brassey's for the International Institute for Strategic Studies, 1992), p. 75.

2 *Guardian*, 13 January 1993.
3 See *Chemical Weapons Convention Bulletin*, 19 March 1993, p. 13.
4 The first initiative to control the export of dual-use chemicals had been taken in April 1984. But the controls and measures imposed by several Western countries were not uniform, either in scope or application. So the Australia Group was set up mainly to harmonize the measures taken individually, and to enhance cooperation among the participating member states on this issue; see the official document entitled 'Current Export Controls on Materials Used in the Manufacture of Chemical and Biological Weapons', published by the Australia Group in December 1992.
5 See *New York Times*, 7 June 1990.
6 See Peter Rudolf, 'Nonproliferation und internationale Exportkontrollpolitik', *Aussenpolitik*, Vol. 42, No. 4, 1991, pp. 395–6.
7 *NRC Handelsblad*, 30 September 1988.
8 See, for a collection of insightful articles: United Nations Institute for Disarmament Research (UNIDIR) Newsletter, No. 20, *Chemical Weapons Convention*, Geneva, December 1992.
9 See Amy E. Smithson, 'Chemical Destruction: The Work Begins', *The Bulletin of the Atomic Scientists*, April 1993, p. 38.
10 See Charles C. Flowerree, 'The Chemical Weapons Convention: A Milestone in International Security', *Arms Control Today*, October 1992, pp. 2–3.
11 The first plenary session of the PrepCom took place in The Hague, 8–12 February 1993; see *OPCW Synthesis*, No. 1, 1 April 1993.
12 See *Convention on the Prohibition of the Development, Production, Stockpiling, and Use of Chemical Weapons and on Their Destruction. Summary and Text.* Washington, DC, Arms Control Association, October 1992.
13 See *Trust and Verify*, No. 35, January/February 1993.
14 See Amy E. Smithson, 'Tottering Toward a Treaty', *The Bulletin of the Atomic Scientists*, July/August 1992, p. 10.
15 Michael Krepon, 'Verifying the Chemical Weapons Convention', *Arms Control Today*, October 1992, p. 23.
16 *Guardian*, 13 January 1993.
17 See James F. Leonard, 'Rolling Back Chemical Proliferation', in *Arms Control Today*, October 1992, p. 16. The Egyptian Foreign Minister, Amre Moussa, argued along similar lines in an interview on 26 December 1992 quoted in *Chemical Weapons Convention Bulletin*, March 1993, p. 14.
18 See on this and related issues: Paul A. D'Agostino, 'Development of a Chemical Agent Detection System (CADS) for Remote Detection of Chemical Warfare Agents in the Persian Gulf', in Peter Brogden and

Walter Dorn, eds, *Controlling the Global Arms Threat* (Ottawa: The Canadian Centre for Arms Control and Disarmament, Aurora Papers 12, July 1992).

19 Figures quoted in Kathleen C. Bailey, 'Problems with a Chemical Weapons Ban', *Orbis*, Vol. 36, No. 2, Spring 1992, p. 245. See also Richard Latter, *An End to Chemical and Biological Weapons?* (London: HMSO, Wilton Park Paper 63, October 1992), p. 5.

20 See Smithson, 'The Work Begins', pp. 39–41. The US Congress has earmarked US$55 million for assisting the destruction of Russia's chemical stockpile.

21 Ibid, p. 40.

22 See, for a cogent expression of this demand, Hassan Mashhadi, 'The Cost of the Chemical Weapons Convention for the Developing Countries', *Disarmament*, Vol. 16, No. 1, 1993, who argued that 'the developing countries maintain that the share of the developed countries will be shifted onto prices of the chemicals they export, so that in the end it will be the developing countries which will have to shoulder the whole burden by importing more expensive chemicals' (p. 82).

23 24 May to 4 June, and 13 to 24 September.

24 See, for an analysis of these problems: Susan Wright, ed., *Preventing a Biological Arms Race* (Cambridge, Mass.: MIT Press, 1990).

25 Quoted in S. J. Lundin and Thomas Stock (with Erhard Geissler), 'Chemical and Biological Warfare and Arms Control Developments in 1991', *SIPRI Yearbook 1992. World Armaments and Disarmament* (New York: Oxford University Press, 1992), p. 179.

26 See Graham S. Pearson, 'Biological Weapons: The British View', in Brad Roberts, ed. *Biological Weapons. Weapons of the Future?* (Washington, DC: The Center for Strategic and International Studies, Significant Issues Series), Vol. XV, No. 1, January 1993.

27 During that period, the USSR exported its Scud and SS-21 systems; the United States exported its MGM-52 Lance systems; China was prepared to sell its SS-2s, and North Korea exported a version of the Soviet Scud. Argentina, Egypt and Iraq were working on their own ballistic missile system (called Condor II), whereas India was developing the Agni and Prithvi missiles, and Pakistan its Hatf 1 and 2 systems.

28 See Richard H. Speier, 'The Missile Technology Control Regime', in Trevor Findlay, ed, *Chemical Weapons and Missile Proliferation: With Implication for the Asia/Pacific Region* (Boulder and London: Lynne Rienner Publishers, 1991), p. 115.

29 In March 1993, Iceland attended the MTCR meeting for the first time (as a non-member).

30 Thomas G. Mahnken, 'Ballistic Missile Proliferation: Seeking Global Solutions to Regional Problems', *Disarmament*, Vol. 15, No. 3, 1991, p. 20.

31 In 1989–90, a West German parliamentary committee of inquiry into the possible diversion of nuclear materials by the German nuclear industry concluded that German firms had a clean record. However, critics have argued that 'The federal government's role (in nuclear exports) has been to ensure that transfers take place as smoothly as possible' (quoted in 'Parliamentary Investigation Concluded: No Proliferation Threat, Germans Say', *Nuclear Fuel*, 12 November 1990). In February 1992, eight managers from three German firms were arrested on suspicion of having exported dual-use technology to Iraq. Exports included both detonators for Scud-B ballistic missiles and dual-use ferritic spacer magnets for the motor stators of gas centrifuges; see 'Eight German Executives Arrested on Dual-Use Exports to Iraq', *Nuclear Fuel*, 2 March 1992. See also Harald Müller, *After the Scandals*, PRI Report, No. 9, Peace Research Institute, Frankfurt, February 1990.

32 Japan has provided US$20 million, the United States and the European Community have allocated US$25 million each.

33 Joint statement issued on 11 March 1992, following a meeting of officials from the United States, the European Community, Japan, and the Russian Federation, about an International Science and Technology Center to be established in Russia. The United States Congress has further allocated another US$25 million from the 1993 military budget to set up the AmeRus Foundation, a non-governmental organization to assist the ISTC, whereas the EC has approved 4 million ecu for scientific cooperation between EC researchers and those of the former USSR (see *Programme for Promoting Nuclear Non-Proliferation (PPNN) Newsbrief*, No. 20, Winter 1992, p. 13).

34 See on this issue: Leonard S. Spector and Virginia Foran, *Preventing Weapons Proliferation. Should the Regimes Be Combined?*, Report of the 33rd Strategy For Peace, US Foreign Policy Conference of the Stanley Foundation, Muscatine, Iowa, October 1992.

35 Ashton Carter, William Perry and John Steinbrunner, *A New Concept of Cooperative Security* (Washington, DC: Brookings Institution, 1992), p. 37.

CHAPTER 6

CONVENTIONAL ARMS: THE DRIVES TO EXPORT

TREVOR TAYLOR

Introduction: the significance of conventional arms transfers

In traditional realist thinking about international relations, military force is the ultimate instrument of foreign policy, ultimate in the senses of being the most costly, the most risky and potentially the most effective. Although there are many doubts expressed about the contemporary utility of the military instrument, military force is still associated particularly with the ability to coerce or compel.[1]

Today it is also conventional wisdom that the capability to prevail in war is closely related to the technology at a state's disposal. The better the weapons and equipment a state possesses, the better its chances in battle, provided that its troops can use the equipment properly.[2] Not surprisingly then, there has long been considerable attention paid to the arms trade, in which states deliberately pass from one to another the means to conduct warfare.

Because of this association of the arms trade with the ability to use deadly force, all states making arms on any scale control their export: legal exports require a licence granted by the exporting government. In a real sense, the arms trade is already controlled by exporting governments. However, the readiness in practice to refuse a licence may be small, and exporting bodies may also seek to export equipment illegally.

The size of the illegal arms trade cannot be measured accurately, although its money value has long been recognized as only a small fraction of the legal trade. This chapter is focused on the legal trade, but it cannot be overlooked that the violent effects of illegal dealings have been considerable since 1945 and could grow in the 1990s. Most people killed in wars since 1945 have been victims of the light arms which are relatively easy to smuggle and which exist in large numbers. Looking forward, should civil wars continue to proliferate, many people will be killed by weapons whose flows neither the United Nations Register of Conventional Arms nor the Stockholm International Peace Research Institute publications[3] even seek to monitor and which do not enter any official arms trade statistics because they have been smuggled.

The legal/illegal distinction in arms transfers can become blurred, as in the case of the Irangate scandal in the US, and there are detailed reports of a quasi-legal trade in which Russian manufacturers sell through Western agents.[4] Another distinction which is less clear in practice than it is in principle is between all arms transfers and the arms trade. The arms trade comprises the international movements of defence goods which are paid for by the customer, whereas arms transfers also include equipment provided as aid. In reality, putting values on items supplied as aid can be a difficult, almost arbitrary exercise. Many arms which are formally traded, for their part, have credit or pricing arrangements which involve a degree of 'aid'.

Modern interest in the arms trade boomed after the First World War, when it was widely perceived that the profits made by the arms manufacturers must have been a contributing factor to the war's outbreak and continuation.[5] In contrast, little was written after the Second World War (in which the US capacity to produce arms played a massive role in the Allies' effort to defeat the Axis states). Moreover, for more than fifteen years after 1945 comparatively few newly produced arms were sold, most arms transferred being supplied without charge under aid programmes. Often they were surplus equipment from the Second World War. By the early 1960s, however, the economic recovery of the states of Western Europe had advanced sufficiently for replacements for Second World War equipment to be bought, and oil wealth meant that some developing states, such as Iran, could afford to buy rather than rely on grant aid. As a reflection of growing Iranian oil income, the United States made a five-year agreement with Iran in 1964 under which, as a US general put it, 'they would buy more equipment from us as their economy became more viable and had a greater capability ... As a result our grant military assistance to Iran will gradually decrease from year to year.'[6] In the event, the Iranian economy grew at a rate of 10.5 per cent a year between 1965 and 1969.[7] As the arms trade grew, so did external interest in its shape and progress.[8] The 1973 oil crisis, after which increased oil revenues led several Middle East states to make unprecedented defence purchases, established the arms business as a focus for governmental, journalistic and academic concern.

Since then consistent attention has been paid to the arms trade, not least because the value of deals is on occasion dramatically high and also because sales have periodically been marked by scandal and corruption.[9] Inevitably, public alarm is greatest when suppliers find themselves in conflicts where they face weapons which they themselves provided, as happened to Britain in 1982 with regard to Argentina and to France in 1991 with Iraq.

This chapter examines the motives of sellers and buyers in the new political circumstances after the Cold War. It looks both at the broad factors at work with regard to three sets of players (the exporting government, the exporting company and the importing state) and at the more specific matters which have to be addressed in arms trade decision-making. First, however, some broad elements in the arms business need to be noted.

Data on arms transfers

Measuring the value and content of even legal conventional arms transfers is extremely difficult. First, there are problems of secrecy to be overcome. Although even without the new UN Register of Conventional Arms, it is hard to conceal for long the transfer of major arms systems such as tanks or combat aircraft, the precise value of deals and exact quantity of equipment transferred can be disguised. In a noted case before the Suez crisis, President Nasser sought to exaggerate to the Egyptian people the number of aircraft received from the Soviet Union by flying the same planes round and round in a military parade in Cairo. Second, there are definitional issues to be addressed. It is not just a matter of deciding what sorts of military equipment should be included, in a spectrum of defence-related goods running from uniforms, through trucks, four-wheel drive vehicles, radios, reconnaissance aircraft and transport helicopters to rifles, fighter aircraft and missiles. There is also the issue of whether to include the value of training packages and other elements such as civil engineering and construction in an arms deal. British Aerospace, as part of its Al Yamama project with Saudi Arabia, which is principally concerned with the sale of aircraft, has undertaken much construction work and provides an extensive English-language teaching service. Finally, attention must always be paid to whether orders or deliveries are being measured. Orders can take several years to be executed and their value can thus be significantly affected by inflation. On the other hand, less equipment may be actually delivered than originally planned in a provisional order and the actual price paid may be negotiated down from that agreed when the order was placed.[10]

Thus data on the arms trade have to be treated with considerable caution. Since 1969 there have been two regular sources of information: the publications of the Stockholm International Peace Research Institute (SIPRI), including its annual *Yearbook of Armaments and Disarmament*, and those of the US Arms Control and Disarmament Agency (ACDA). However, the latter's *World Military Expenditures and Arms Transfers* has not appeared since the end of 1991 and its future must be uncertain. SIPRI trade data cover only major conventional weapons systems. Neither SIPRI nor ACDA can be seen as producing accurate measurements of the total trade on conventional defence equipment, but both offer time series of data helpful for the trends which they expose.

Using information collected by the US government, the American Congressional Research Service (CRS) has also prepared major reports on arms transfers to developing countries, including assessments and detailed analyses of the value of agreements and deliveries. Perhaps because it has some access to classified information, the CRS produces data which indicate higher values of deliveries than do the figures from ACDA.[11]

In addition, national governments prepare and publicize estimates of their own arms exports values which often greatly exceed the figures suggested by SIPRI or ACDA. For instance, the British Defence Minister responsible for procurement, Jonathan Aitken, told parliament on 9 February 1993 that British arms exports in 1992 were worth £4.2 billion, 20 per cent of the world market. Government figures, however, normally include contracts on such items as the construction of military housing.

As the last year of ACDA data is 1989, Table 6.1 (at end of chapter) uses SIPRI and CRS data. Section 1 of the table shows how markedly defence exports have dropped since the end of the Cold War, in part because the exchanges among members of the former Warsaw Treaty Organization have disappeared. Not shown is the fact that arms deliveries in earlier years were at a higher level than in 1988, sometimes by a considerable amount, not least because of the Iran–Iraq war: the CRS finds that arms deliveries in 1984 reached $55.8 billion at 1991 prices, of which $37.8 billion were supplied by the USSR.

The drop in supplies to the developing world is formidable, as Sections 2 and 3 of Table 6.1 show. The drop in demand is marked but less substantial when, as in Sections 5 and 6, allowance is made for the fact that Russia has stopped supplying arms as military aid. Overall, with the end of the Cold War the proportion of total exports going to industrialized states did not fall, indeed it rose from 41 per cent in 1988 to 49 per cent in 1992.[12]

Sections 6 and 7 of Table 6.1, which look at orders, signal that the fall in commercial demand may have levelled off, particularly when Russia is excluded. The boom in orders in 1990, caused by the Kuwait crisis, led to a fall in 1991 but overall demand seemed to be holding up at a global level of about $31 billion a year.

Arms are exported by many states but six dominate the developing-country market, accounting for around 90 per cent of all sales (see Table 6.2). Table 6.2 shows that SIPRI and the CRS disagree on the Soviet/Russian share of that market, with SIPRI noting minimal Russian sales in 1991. SIPRI points to the United States and China having a bigger share of the developing-country market than the CRS would suggest, while the CRS, unlike SIPRI, sees Britain enjoying a larger market share than France. Of course the two sources do measure different things, with SIPRI dealing only with major items. It is also not clear whether the CRS treats Israel as a recipient, that is a developing country, which may help to explain its lower estimate of American transfers. Both the CRS and SIPRI point to China as a significant player but the CRS sees its share diminishing while SIPRI argues for rising Chinese importance. Not yet apparent in the published data is Germany's role as a significant defence exporter in 1992 because the Conventional Forces in Europe Treaty required it to cut its stocks and it is providing a considerable amount as aid to less well-equipped NATO allies. It is also trying to get rid of much East German equipment which it inherited; for instance, it has concluded a major naval deal with Indonesia.[13]

Tables 6.3 and 6.4 give information on arms recipients. The prominence of Middle East countries in the world market, let alone the developing-country segment, is apparent. In the developing world alone, according to SIPRI, seven of the fifteen largest importers of arms are located in the Middle East, where reduced deliveries to Iraq (under a UN embargo since 1990) and Saudi Arabia are part of a pattern of falling imports. This picture reflects the relatively low price of oil and the end of the eight-year Iran–Iraq war in 1988. However, this trend may well not continue long because of Iran's efforts to build up its forces and the significant Middle East orders taken by the US alone since the end of the Gulf war. Some of these orders, however, have already been reduced so that, for instance, in the spring of 1993 Greece was negotiating to buy 20 Multiple Launch Rocket Systems ordered by the Saudis after the Kuwait crisis which they then found they could not afford.[14] In the Far East, Japan, South and North Korea, and (not in the tables) Taiwan[15] are significant importers whose purchases were falling in the early 1990s. The end of the Cold War (and the UN Security

Council cooperation involved) has meant that some previously significant recipients such as Afghanistan, Angola, Cuba, Iraq, North Korea and Vietnam have lost their importance.[16] In the main these were aid recipients and political allies of the former Soviet Union.

With this qualified note on the overall pattern of arms transfers, attention can be turned to the motivations of the participating entities, both governments and companies.

The motives of exporting governments

There are often economic, foreign policy and national strategic concerns favouring arms exports. The weight of any single consideration clearly varies from case to case.

Economic considerations

Economically, arms exports, like any other exports, should generate employment and foreign exchange. Table 6.5 gives some sense of what is at stake in the major arms-exporting states by listing estimates of defence industrial employment in total. The share of total defence production provided by exports is difficult to assess and varies from year to year, but in round terms West European countries have exported about 40 per cent of their defence production while the US has exported about 15 per cent.

The significance of employment considerations is heightened by the fact that major defence systems are often produced in large plants whose work accounts for a large share of employment in the community.[17] Russia has several cities devoted almost entirely to defence production. The loss of such large employers cannot easily be offset by new economic activity, either in the new Russia which is struggling to introduce economic reform and to halt economic decline, or in the West where economic recession became pronounced just as the Cold War was ending.

Even for the major suppliers, as Table 6.6 makes clear, arms exports are but a small part of total exports. However, it is also salient that four of the top six arms exporters have significant balance-of-trade deficits and are thus particularly concerned to maximize their foreign exchange revenue.

One factor modifying the impact of foreign exchange and employment considerations for exporters is that arms purchasers predictably try to minimize their economic costs. They attempt to reduce the foreign exchange which they pay out and to locate some or most of the employment concerned with a purchase within their own state. The two most obvious core ways of achieving these ends are either to arrange to produce as much as possible of a system under licence in the purchasing state (as Japan has done with its F.15s), or to arrange an offset package under which the arms-selling state or company also agrees to buy goods from the purchasing country. The significance of licenced production can be seen in the fact that, although Japan deploys many 'US' systems and spends much more on defence than Greece or Turkey, in 1991 and 1992 these two Mediterranean states emerged as bigger arms importers than Japan.[18]

There are many variants on these approaches: offsets can be associated with coproduction (under which a buying state will produce all of some specified components for a particular system. The components are shipped to the selling company for final assembly and sale either to the main manufacturing state's own armed forces or some other customer.) In 1993 the four European states which together bought the F.16 in 1976 (Belgium, Netherlands, Denmark and Norway) are negotiating an 80 per cent offset against the costs of their mid-life update for the aircraft. They want the offset to include the building of 15 per cent by value of all F.16s manufactured, a concession which they had won when they bought the aircraft. The US is resisting this coproduction demand.[19] Offsets may specify that some other defence equipment is to be bought from the main arms importer, or quite unrelated civilian items may be involved. Sweden, negotiating a tank purchase in 1993, wanted a deal which would enable it in future to build its own tanks. Giat, trying to sell the Leclerc, was already marketing from France the Swedish CV.90 armoured vehicle family and has said it will build all Leclerc armoured recovery vehicles in Sweden if it receives the Stockholm tank order. Rival manufacturers are offering other packages.[20] The British offset deal on the seven AWACs which it ordered in 1986 specified that 'high technology' work should be given to British firms. Clearly, offsets can merge into barter arrangements (the Saudis are paying in oil for their Al Yamama purchases).

Offsets are frequently unpopular with exporting governments and firms, not least because they are difficult to administer and often cause arguments between purchaser and seller. Nonetheless they remain appealing to buyers, and the efforts, mainly by the US, to outlaw offset deals in intra-NATO trade have been unsuccessful. Licenced production and coproduction clearly help to spread defence manufacturing capabilities to new firms and states, so increasing the prospect that national production might substitute for trade in the future.

The foreign exchange benefits of arms trading can also be much modified by the extended and generous credit arrangements which often have to be made to finalize sales. Moreover, the purchaser may default on payments, as Egypt did when it broke with the USSR in the mid-1970s. As a consequence of confronting Iraq over Kuwait, France lost all hope of being paid the $6 billion or so which Iraq owed it for arms supplied during the Iraq–Iran war. A major reason for believing that Russia will struggle to complete major arms deals in future is that it will find it hard to provide appropriate finance. Arranging finance for a modest Russian–Turkish deal for armoured vehicles has already proved difficult.[21] Cash deals for major systems are rare.

Yet, despite these problems, arms exports remain economically attractive to many governments. Indeed most NATO states seem explicitly to want to avoid a balance-of-trade deficit on defence. If, like Norway, they need to import because their range of national production is small, they like also to export some of the defence equipment which they can make themselves.

Overall, the foreign exchange and employment benefits of arms sales would appear least significant to those countries with growing economies, limited unemployment and a balance-of-trade surplus. They are most valuable to those with struggling economies, and Russia's current enthusiasm for arms sales needs no further emphasis:[22] there are real concerns that it will be ready to sell even advanced ballistic missile technology, particularly to China.[23] But the end of the Cold War has been accompanied by recession in the West and arms exports have been a valued activity everywhere: by the end of the Bush administration the US seemed reluctant to allow any other suppliers to conclude sales in the Arab Gulf.[24] According to US Commerce Department figures, the US trade deficit grew by $19 billion to $84 billion in 1992, but the surplus on defence goods was $9.6 billion.[25]

A further economic aspect of arms sales is that they can be used to recoup some of the R&D costs which were sunk by the manufacturing state into the system's development; exports thus enable that state to pay a lower price for the equipment bought for its own forces. This point needs at least brief reference to cost structures:

(1) Because of the rising technological content of weapons, the post-1945 period has seen a tendency for increases in the fixed costs (R&D and production investment) associated with successor generation of equipment types. In the UK equipment budget, R&D spending is about half production spending, indicating that, for many British systems, the fixed costs of development and production investment per unit are about half the actual manufacturing cost.
(2) In addition, defence production, like other manufacturing, is marked by learning curves, so that as more of a system is made, costs fall in proportion. A rule of thumb in the aircraft industry has been that, as output doubles, so the unit cost of production should fall by perhaps 20 per cent (if the first aircraft on a line costs $100 million to assemble, the second should cost $80 million and the fourth $64 million).[26] As manufacturing becomes more 'computerized', however, shallower curves may prevail.
(3) There are also often economies of scale in defence production, so that a lower unit cost should be obtained by making, say, eight units per week rather than four. Economies of scale clearly vary with the equipment under consideration and there are some grounds for believing that efficient low-rate production may be possible using modern, flexible manufacturing methods.

The actual economic benefits of a particular arms sale depend on the prices charged, the timing of production and the distribution of the benefits between the manufacturing firm and its home government. It is enough here to point to the principal factors and relationships at stake:

- arms exporters, depending on their bargaining power, may or may not feel able to charge the purchaser an R&D levy, perhaps 7.5 per cent of the manufacturing cost, as a contribution towards the original fixed costs of the project. Such a levy may be paid to the company or the exporting government, depending on who first paid for the R&D;
- arms exporters may charge a customer the marginal or the average cost of production. If the customer pays the marginal cost of production, and if the export order is completed after the home market has been satisfied, the firm will have proceeded well down a learning curve and the marginal costs will be much lower than they were for the home government's purchases;

- thus the optimum saving for the exporting government occurs if the recipient government pays to the exporting government an appropriate R&D levy and also pays the average production price (or more) to the company. The savings can be transmitted to the exporting government by having it pay a lower price for the systems it buys for its own forces compared with the price it would have had to pay without exports. For instance, it is likely that the cost to the US Air Force of its AWACs fleet has been much reduced by sales of the aircraft to Japan, NATO allies and Saudi Arabia.

The above is what can happen in principle. The secrecy of financial aspects of the arms trade means that demonstrating what is happening in practice is almost impossible. There are cases where the R&D levy paid by importing states clearly benefited the exporting government[27] but often the bargaining power of the purchaser is such that prices can be driven down in an international competition.

Despite all the doubts and difficulties, arms exporting remains an attractive economic option for those states which wish to sustain a defence industrial capability and whose economies are not awash with employment opportunities and foreign exchange. Moreover, the economic benefit from sales of major systems comes not just from the revenue of the hardware initially delivered. Of equal if not greater value is the gain from contracts on related services, such as construction and training, and from contracts during the life of the system for spare parts and maintenance.

Foreign policy considerations

Arms exports can also be seen as supportive of foreign policy. During the Cold War when the US was seeking to contain Soviet power, it used arms exports, often supplied as aid, to boost its allies. The trauma of the Vietnam war led to the promulgation in 1969 of the Nixon Doctrine, under which the United States agreed to supply arms rather than US forces to its friends under pressure. Since the mid-1960s the United States has supplied Israel generously with arms, in part so that US forces themselves would never have to fight to save Israel.

But supplying arms can also send a signal that the supplier would if necessary come to the recipient's help. Certainly this is the hope of recipients such as the Saudis, who buy from those states whose security and defence support they value. Sometimes readiness to buy arms from a state is a tacit condition of the supplier's support. Some would judge that Washington might lose enthusiasm for defending Japan's security were the United States not Japan's preferred arms supplier.

Arms exports in these senses are an aspect of the capacity of a state to defend its interests either by deterrence or in the event of attack, although the defence is executed by others. However, to enhance the capability of a friendly state to fight, arms exporters must frequently devote considerable effort to training recipients in the maintenance and use of equipment, both in isolation and as an element in wider military tactics and operations. The USSR failed to take such steps when it supplied Egypt with a mass of tanks and aircraft from 1955; as a result Israel could attack and quickly rout Egypt in 1956, capturing much Soviet equipment intact in the process. More than 30 years later, Kuwait's suppliers clearly failed to instil in Kuwaiti forces the readiness/ability to fight when Iraq invaded in 1990.

Arms exports can also be seen as a foreign policy tool to change the general or particular behaviour of the recipient, as an inducement for good behaviour. Here arms transfers have a limited reputation for success. The post-1945 period is marked by states using arms for purposes of which the suppliers did not approve. The United States objected to Turkey's use of US arms in the invasion of Cyprus in 1974 and to Israel's invasion of Lebanon in 1982. A long-standing arms relationship with Moscow did not prevent Egypt turning to the West in the mid-1970s. The USSR reacted negatively to the behaviour of its close client Iraq for the attack on Iran in 1980, and broke with Somalia after it used Soviet arms to invade the Ogaden in the 1980s. Neither of Iraq's main suppliers in 1990, the USSR and France, approved of its invasion of Kuwait. Most recently Germany has strongly resented Turkey's use of German-supplied equipment in the war against Kurdish rebels.

A case of particular significance is that of Israel. US readiness to supply arms in abundance to Israel, particularly since President Johnson agreed to supply F.4 Phantoms, in part reflects the US belief that Israel will not make the concessions needed for peace unless it feels it has a reliable source of arms. Indeed, without reliable conventional arms supplies, the United States sees an increased danger of Israel adopting an overtly nuclear stance. Yet this broad approach cannot be seen as a success. Significantly, Israel is widely acknowledged to have tens, if not hundreds, of nuclear warheads.[28] On the other hand, withholding arms has apparently swayed Israel: in

1975–6 the United States did briefly but successfully exercise pressure by restricting arms deliveries, and so forced Israel to accept the Sinai II agreement.

The limited influence of arms exporters on their customers' wider behaviour is predictable: the world arms market is still a buyer's market, and states importing arms understandably want to minimize the extent to which they compromise their freedom by their procurement actions. States which can afford to procure from a range of suppliers often do so, avoiding dependency on any single one, even though there may be some direct military disadvantages involving loss of standardization and interoperability. Also, Israel has sought to minimize its dependence in crisis on even the United States by building up its own conventional as well as nuclear industry. The manufacture of ammunition and spare parts used in high volume has been of particular importance.

As noted earlier, arms export policy can be seen to have failed when a supplier finds itself having to fight against its own equipment in the hands of an enemy. Certainly this consideration serves to underline the fact that major military equipment should be sold only after the most careful deliberation as to the purposes and aspirations of the recipient. However, practical considerations and risks must be recognized. A fundamental point is that weapons often have a protracted life whereas political circumstances can change very quickly, as events in Iran in 1979 showed. Moreover, in the 1980s the West had a real interest in Iraq's avoiding defeat at the hands of Iran and it is not surprising that Baghdad received Western military help. Further, to refuse to sell a state arms which it feels it needs for its legitimate self-defence can be expected to be seen as an unfriendly act, increasing the hostility of the would-be recipient towards the reluctant supplier. Nasser's nationalization of the Suez Canal Company in 1956, which led to the Suez war, was due in large part to the refusal of the Western states to supply Egypt with even modest amounts of arms.[29]

In some senses the end of the Cold War clearly reduces the potential of the foreign policy gains from arms exports. Since East and West are no longer struggling on a global scale for influence, they will be less concerned about how certain states behave. This consideration has clearly meant that Russia in particular has become much less concerned to supply arms for other than hard currency; indeed the potential gains from arms exports seem to be viewed by Moscow in 1993 as solely economic.

However, the longer term may throw up different pictures. Major states, without the stakes of the Cold War, may become ever more reluctant to deploy their forces overseas to protect their interests, and ever more ready to equip the forces of others to do the same job. The clear instinct of the Clinton administration on Bosnia has been to lift the arms embargo, and indeed it may be some time before the UN again so hastily imposes an arms ban on all parties in a conflict. Recipients, from their perspective, may see the protection of great powers' forces as ever more unreliable after the Cold War and so could well seek to build up their own strength by arms imports. It is widely acknowledged that the existing tendencies to acquire arms in the Far East would be much reinforced were the United States to withdraw from the region. The competition in arms acquisitions in the Gulf dates, after all, not from the oil price increase of 1973, but from the UK decision in 1968 to withdraw from the Gulf the very modest forces with which it supposedly policed the area. The final consequences of the end of the Cold War for arms supplies as a foreign policy instrument have not yet been seen.

Strategic considerations

The strategic motivations for arms exports concern the role that such exports can play in enabling a state to sustain its own defence industry, and enjoy the concomitant perceived security and autonomy.

A prime consideration here is that, although the rate of technological change in defence equipment can be very fast, much defence equipment lasts a very long time without wearing out. Big systems like combat aircraft, tanks and warships are expected to be in service for 20 to 30 years. National armed forces, on the other hand, usually want a new generation of equipment to be introduced over a much shorter period, perhaps five years. This raises the question of what factories and development teams do once national markets have been satisfied. Export is the obvious answer until new home orders come along.

This problem is growing in seriousness. Even during the Cold War, because of rising costs, armed forces in the developed states were pressed to procure a new generation of a system in smaller numbers than its predecessor. With the Cold War over, defence equipment spending in NATO and the states of the former Warsaw Treaty Organization has fallen drastically. Russian procurement spending went down by about 60 per cent in 1992. Table 6.7 illustrates that in NATO equipment spending has decreased much more sharply than defence spending as a whole.

Looking forward, some defence manufacturing prospects could be eased by the development of computer-integrated manufacturing and other industrial advances which will make possible efficient low-rate production and will add flexibility. Plants which manufacture tank subsystems may be able to adapt to making parts for warships, or even civil goods.

In 1992, US Secretary of Defense Les Aspin advanced a four-point scheme for the maintenance of the defence industrial base in the United States which did not involve exports. He advocated periodic upgrading of major systems and low-rate production of selected new items to sustain production capabilities. In addition the United States would press ahead with R&D and make extensive use of prototypes without normally building the new technology into produced systems. Technology, he argued, could be kept on the shelf for use when needed. Finally he recognized that the United States should press ahead with those specified 'silver bullet' advanced systems in such areas as ballistic missile defence where the country would gain a decisive military advantage from the systems involved.[30]

However, the practicalities of this scheme remain to be demonstrated and in the meantime it is striking that senior US air force and army personnel have recognized the value of exports as a means of keeping open US supply lines for tanks and aircraft. Despite the liberating effects of flexible, computer-integrated manufacturing, it seems unlikely that any state will be able to sustain a broad-based advanced defence industry without some access to international markets, either through collaborative development and production projects, widely used in Western Europe, or through exports. Because of the limitations of the national market, an increasing number of states, as diverse as India, Iran, Slovakia, South Africa and Poland, want to export more arms.[31] Significantly, by the beginning of 1993 France was ready to risk its extensive trade and investment relations with China by selling 60 Mirage 2000 aircraft to Taiwan,[32] and the United States risked loss of Chinese cooperation on non-proliferation issues by selling F.16s to the same country. The aircraft industries of both France and the US needed the orders, not just for profit, but for survival.

While the above factors broadly press governments to allow, even encourage arms exports, it remains the case that many specific issues have to be addressed in practice. No major government allows significant equipment to be exported without giving every case careful consideration, and states vary considerably, for instance in the support which they give to their companies seeking to conclude export deals, and in the precise shape of their arms export policies.[33] At Maastricht the states of the European

Community committed themselves to developing a common policy on arms exports, a target which will be achieved only with considerable difficulty, given their currently rather varied assessments of the advantages of arms exports.

The arms exporting company

Many of the considerations which drive an arms-exporting government also affect the companies concerned. Most firms recognize that they need to export to survive and will probably need exports more in the future as national markets shrink.

A basic characteristic of companies is their commercial nature: they need to sell to generate turnover and hopefully profit. But there is a fundamental question as to why defence development and production is so widely entrusted to such commercial concerns.

Historically a spectrum of organizational arrangements has evolved for the development and production of arms. At one extreme is the privately owned company, perhaps producing both civil and military goods. Some such companies are very large, for example the German Daimler Benz group which includes DASA, Germany's main aerospace firm. At the other is the state arsenal, owned by the state and organized as an administrative support for the armed forces. Such bodies operate within a budget but they are not expected to make a profit, being an arm of the civil service. Until the early 1980s the Royal Ordnance factories in Britain were of this type. The United States still operates some government-owned arsenals. Between the two extremes are private companies which operate state-owned facilities such as Devonport Dockyard in the UK and a range of plants in the United States. There are also state-owned enterprises such as Aerospatiale in France, which are expected to compete in world markets like ordinary companies but which normally must raise their capital from the state rather than the private sector. Sometimes, as in Italy, the state owns defence firms through an intermediate holding company.

The important message as far as arms exports are concerned is that increasingly states have found that the organizations which perform best in the development and production of defence equipment are those which, because of exposure to competition, have real incentives for effective and innovative performance.

In defence, international markets are of particular importance as far as competition is concerned because increasingly national markets in Europe and even the United States are dominated by only two suppliers or even one. Effective domestic competition is therefore difficult. France has led the way in establishing national champion companies, such as Aerospatiale (transport aircraft, helicopters and ballistic missiles), Dassault (combat aircraft), Thomson-CSF (defence electronics) and Snecma (jet engines); these have a monopoly at home but to survive they need to export. Most such companies have been state-controlled. In the 1970s and early 1980s such bodies served France well, but since the late 1980s their export success has moderated considerably.

The important point then is that arms-producing states have seen that the most effective way to procure their equipment is often from a commercial enterprise; they have moved away from the civil service-based arsenal. Philip Noel-Baker's plea in the 1930s for government ownership of the defence industry, partly on the grounds that private ownership brought few benefits, has been widely rejected.[34] Commercial organizations are often interested in increasing sales through exports, and governments may judge that enterprises which can survive in international defence markets should also serve their home armed forces well. However, there may be different perceptions between companies and government as to the sort of equipment which should be developed, with companies preferring that which would yield maximum global sales while the home government wants the equipment of most value to its armed forces. Different states have approached this matter in contrasting ways. In the United States, where exports have generally accounted for only about 15 per cent of defence manufacturing because the domestic defence market has been so large, the export possibilities of equipment have rarely been significant. When Northrop attempted to develop and sell the F.20 as a successor to the F.5 as a 'simple' combat aircraft, the US Air Force would not buy the plane and foreign buyers were not interested in a model which the USAF did not endorse by a purchase. The United States has often wanted to keep the latest version of a system for itself and to export only an earlier version.

British policy has been rather similar in practice to that of the United States: although export potential has long been a factor of which the Ministry of Defence is supposed to take account in drawing up the requirements of a system, in practice it is difficult to find cases where the UK requirement has been compromised to take account of export possibilities. On the other hand, Britain has exported more modern versions of kit than its own armed forces possessed, notably with the sale to Iran of Chobham armour-equipped Chieftain tanks in the 1970s.

In France, on the other hand, the combination of arguments that the basis of French defence was the nuclear arm and that strategic autonomy should be pursued through a self-sufficient armaments sector meant that export potential has often seemed, at least from outside, to be a major influence on the equipment procured for French forces.

In the Cold War, Soviet equipment was often less sophisticated technologically than its Western equivalents, which in some ways made it appealing to developing-country purchasers, but nevertheless Moscow often exported less advanced versions of equipment than its own forces possessed. Crucially, for many Soviet exports, commercial considerations were secondary; indeed in the 1980s the USSR may have exported as much as $80 billion worth of military equipment for which it did not receive payment.[35]

From a corporate perspective the defence export trade raises issues comparable to many of those in civil commerce. For instance, how many resources should be devoted to marketing and what forms of marketing should be allowed? Completing the sale of a major piece of defence equipment can take five years or more of negotiations and obviously that can involve the expenditure of hundreds of thousands, if not millions, of dollars by the company. Moreover, the need often to pay commission to local agents associated with governmental procurement presents companies with ethical and legal issues over how funds may be used.

Firms contemplating defence exports also frequently have to negotiate the credit arrangements to be offered, the terms of payments, any offset and coproduction arrangements and technology transfer matters. In all these issues they frequently need to work closely with their governments as well as with their potential customers. The complexity of the arms trade is further apparent when account is taken of coproduction arrangements, which may well mean that the main exporting company has to get its own government's permission to transfer technology, and the cooperation of its own subcontractors at home to arrange the technology transfer in practice.

Thus it may be concluded that arms exporting for companies is often a less attractive option than selling to a perhaps sympathetic home government with which close contact can be easily maintained, and which usually pays its invoices on schedule. Exporting in the defence business is something which companies attempt when they need to. As national defence budgets shrink in the West and in Russia, however, export markets look more and more appealing to companies desperate for orders.

Traditionally the major exporting companies work closely with their governments to win sales, although from the Carter to the Bush administration the US government was less supportive of arms exports. In the early 1990s, however, governments and company interests began in some cases to diverge, and the end of the Cold War has left many governments with surplus equipment which they are not obliged to destroy under the CFE Treaty but which they do not wish to keep. By releasing such equipment onto the international market, often at low prices, governments are cutting into the market for new equipment.[36]

Overall the wider picture is not encouraging for defence firms in Russia and the United States, or indeed Western Europe, since there seems no way that developing-country orders can offset the major decreases in the massive amounts which the superpowers previously devoted to defence. Even at 1988 prices, US equipment spending in that year, using the NATO definition, was $72.7 billion, almost twice the value of all CRS-measured arms deliveries to the developing world in the same year. US firms, which relied on the home markets for 85 per cent of their sales, must more than double their exports to fill the gap left by even a 20 per cent drop in domestic orders. European firms, which traditionally export 40 per cent or more of their production, need more marginal adjustments to their exports to offset declining home sales, but they now face much fiercer competition from larger US firms and will continue to do so. Unless West European firms can organize themselves into effective units which can serve their home markets efficiently and win contracts overseas, the turn of the century could leave the US defence industry dominant in the world, although it too will be forced to reduce drastically in size.

Importing governments

A final perspective on the conventional arms trade is that of the importing state – the demand side of the equation. It is easy to overlook the fact that arms transfers occur primarily because of the perceived needs of the purchasers, not the economic or other needs of the exporters. Clearly, this is not to say that those buying arms necessarily choose wisely.

Outside the area of former Cold War confrontation in Europe and North America, it is difficult to discern at the macro level why demand for arms should decline over the longer term, unless the prospects for cooperative regional and global order improve significantly or economic growth in currently successful areas declines. This applies particularly to two regions addressed more fully in an earlier volume in this series.[37]

In the Far East, states increasingly want air and naval forces as well as ground-based technology to patrol and monitor their air and sea space. The growth of Chinese power projection capabilities is likely to worry many ASEAN states involved in the Spratly and Paracel island disputes (as well as Japan). The effort to build more structured cooperative security relations among Far Eastern states, particularly by strengthening international bodies such as APEC and the ASEAN Post-Ministerial Conference, is not yet assured of success. Historical rivalries, suspicions and fears could yet motivate arms acquisition by economically successful states with much to lose.

In the Middle East the balance of power in the Gulf is inherently problematic. In order to match the population and wealth of Iran, Iraq needs the support of Saudi Arabia, but a cooperative relationship between them is impossible given the contrasts between the governments of the two states. Saudi Arabia has massive wealth but extensive borders which are hard to defend with a small population. The Arab sheikhdoms constitute small states which can quickly be overrun, as Kuwait found to its cost. A stable relationship of conventional forces is thus hard to envisage. Elsewhere in the Middle East, progress towards an Arab–Israeli settlement remains uncertain and, even in the event of a peace agreement, Israel's vulnerability would lead it to seek always to be able to conquer its neighbours quickly. Although there is periodic talk of a Conference on Security and Cooperation in the Mediterranean to help bring cooperative security to the Middle East and North Africa, the prospects for building effective multilateral security arrangements in this region appear limited.

South Asia has no obvious equilibrium. The capabilities which India is motivated to acquire in order to match China will always alarm Pakistan and the significance of the religious divide between India and Pakistan shows no sign of diminishing. When the South Asian region as a whole is considered, there seems little practical alternative to Indian domination, a position which India's neighbours will surely resist.

The data on arms transfers noted earlier show that the flow of arms is strongest where there is a combination of serious disputes and wealth, as in the Middle East. Disputes among poor parties are less likely to generate major arms transfers after the Cold War, as the drying up of deliveries to North Korea and Afghanistan signals, and in such areas ground forces with light arms and artillery are likely to predominate. Overall, it is true that the current level of arms demand is diminishing in terms of value. Moreover, the admittedly rather dated ACDA information shows that states in the developing world and in the world as a whole are tending to devote less of their defence budgets to arms imports, perhaps because of increased domestic arms production. Average annual world arms imports in the 1986–9 period were up only 8 per cent on their 1979 level, whereas world military expenditure was up by 24 per cent. Among developing states, annual arms imports were down by 2 per cent in the 1986–9 period compared with 1979, while military expenditure was up by 9 per cent. None of this is incompatible with many state and non-state groups in dispute resorting to organized violence to achieve their aims. Also, it should not be overlooked that the arms trade is to a degree cyclical as states concentrate on absorbing what they have bought before settling on further acquisitions.

Finally, any analysis aimed at explaining the arms trade should not neglect the difficulties which a recipient faces in trying to decide what should be acquired, and the consequent criticisms which are often levelled against individual selections. Defence procurement is a problematic task for all governments and indeed no government, whether American, French, Japanese, Russian or Chilean, can be held up as a model of excellence in this respect. This is not surprising since governments which themselves arrange the development of equipment must estimate (or guess) the answers to several questions: what characteristics should a piece of equipment have to make it militarily useful in perhaps 30 years time; what will those characteristics cost to develop and operate (two very different calculations); and which items should have priority given that funding is limited?

States simply buying equipment 'off the shelf' rather than arranging its development must execute all these tasks except the estimate of development costs, yet such states are often located in areas of great political instability where the possible future tasks of their armed forces are difficult to foresee with confidence. Moreover, many developing (and some developed) states[38] lack the technical expertise to evaluate the relative merits of the different offers which they may receive: the documentation involved is massive and complex for any significant system. Armed forces everywhere place a heavy stress on symbolism and therefore it is also predictable that they will often be drawn to items which they see as conferring prestige on them and their country. Finally, in countries which lack an established tradition of a loyal civil service diligently and objectively serving the state, it is difficult to ensure that those responsible will not choose for their armed forces that system which brings them the most personal benefit.

In summary, for the purchasing government, arms importing issues are only an element in the broader problems of defence acquisition and procurement.

Conclusion

This chapter has pointed to the economic, foreign policy and strategic reasons why states transfer arms, why companies want to export, despite the difficulties involved, why states are increasingly entrusting arms development and manufacture to commercial organizations with a business ethic rather than to civil service-oriented state organizations, and why states want to acquire arms but face real problems in deciding just what to acquire. These essentially are the forces which need to be managed and channelled if multilateral controls on arms transfers, as opposed to existing national controls, are to be developed.

Looking forward, the foreign policy reasons for arms transfers have been much reduced by the end of the Cold War, but the relative economic value of arms exports to selling states and companies has been enhanced by falling home defence markets, and the strategic value of arms exports as a means of preserving a defence industrial base is also rising. Arms exports cannot and should not prevent a significant shrinkage in the human and physical resources devoted to military industry in the West and the former Soviet bloc, but they may enable a state or group of states to sustain continuous development and manufacturing work in key areas and to facilitate the expansion of important production in an emergency. This latter factor is perceived as particularly important in the United States and, if Western Europe does not undertake some collective thinking about the sorts of defence industrial capability which it wants to preserve for the longer term, the US could well come to enjoy a *de facto* monopoly in most areas of advanced defence technology. Whether or not that would matter is, of course, a subject for another work.

Table 6.1: Data on trends in arms export values (in US$million)

	1988	1989	1990	1991	1992
Section 1					
Total exports of major conventional weapons at 1990 prices (SIPRI)	40,034	38,133	29,972	24,470	18,405
Index	100	95.3	74.9	61.1	46.0
Section 2					
Total exports of major conventional weapons to developing states at 1990 prices (SIPRI)	23,688	21,623	17,682	13,240	9,320
Index	100	91.3	74.6	55.9	39.3
Section 3					
Total arms deliveries to developing states at 1991 prices (CRS)	43,796	35,345	30,624	18,365	—
Index	100	80.7	69.9	41.9	—
Section 4					
Exports of major conventional weapons to developing states, excluding Russia/USSR exports, at 1990 prices (SIPRI)	13,408	11,275	11,087	9,433	7,416
Index	100	84.1	82.7	70.3	55.3
Section 5					
Arms deliveries to developing states, excluding Russia/USSR exports, at 1991 prices (CRS)	22,218	16,538	17,386	11,965	—
Index	100	74.4	78.3	53.9	—
Section 6					
Total arms transfer agreements with developing states at 1991 prices (CRS)	30,933	30,610	41,109	24,661	—
Index	100	100	132.9	60.0	—
Section 7					
Total arms transfer agreements with developing states, less agreements with Russia/USSR, at 1991 prices (CRS)	18,433	19,110	29,909	19,661	—
Index	100	103.7	162.3	106	—

Sources: R.F. Grimmett, *Conventional Arms Transfers to the Third World 1984–1991*, Washington, DC, Congressional Research Service (CRS), 1992; SIPRI, *SIPRI Yearbook 1992: World Armaments and Disarmament* (Oxford: Oxford University Press, 1992).

Table 6.2: The major arms exporters to the developing world (Deliveries value in 1991 US$million)

	1988	1989	1990	1991	1992
United States					
Deliveries value (CRS)	5,136	3,784	5,438	5,365	
% share of total (CRS)	11.7	10.7	17.8	29.2	
% share of total (SIPRI)	19.0	16.9	26.1	31.3	33.0
Russia/USSR					
Deliveries value (CRS)	21,578	18,807	13,328	6,400	
% share of total (CRS)	49.27	53.2	43.5	34.9	
% share of total (SIPRI)	43.4	48.2	37.4	4.1	1.5
France					
Deliveries value (CRS)	1,124	1,189	3,883	1,000	
% share of total (CRS)	2.6	3.4	12.7	5.5	
% share of total (SIPRI)	7.0	9.5	10.1	5.5	3.8
United Kingdom					
Deliveries value (CRS)	4,046	4,323	3,883	3,000	
% share of total (CRS)	9.2	12.2	12.7	16.3	
% share of total (SIPRI)	6.4	9.2	6.6	5.2	7.1
Germany (West)					
Deliveries value (CRS)	674	324	105	600	
% share of total (CRS)	1.5	1.5	0.3	3.3	
% share of total (SIPRI)	1.2	1.0	4.8	3.2	3.2
China					
Deliveries value (CRS)	3,259	2,378	1,469	900	
% share of total (CRS)	7.4	6.7	4.8	3.3	
% share of total (SIPRI)	8.6	4.4	7.1	12.9	16.5
All six states					
Deliveries value (CRS)	35,817	30,805	28,106	17,265	
% share of total (CRS)	81.8	87.2	91.8	94.0	
% share of total (SIPRI)	85.8	88.8	92.2	88.3	83.9

Source: R.F. Grimmett, and *SIPRI Yearbook*, op. cit.

Table 6.3: Major recipients of arms: all countries (in 1991 US$million)

	1988	1992	1988–92
India	3,709	1,197	12,235
Japan	2,544	1,095	9,244
Saudi Arabia	2,441	883	8,690
Afghanistan	1,264	0	7,515
Greece	814	1,918	6,197
Turkey	1,447	1,511	6,167
Iraq	2,845	0	4,967
Germany	514	144	4,473
Spain	1,653	370	3,747
Iran	648	877	3,632
South Korea	1,125	414	3,524
Czechoslovakia	1,122	4	3,501
Pakistan	334	1,000	3,486
USSR/Russia	1,421	0	3,383
Thailand	518	869	3,271
Others	17,656	8,693	67,003

Source: *SIPRI Yearbook*, op. cit.

Table 6.4: Major Third World recipients

Country	Deliveries* (1988–91)	Country	Deliveries* (1988–91)
Saudi Arabia	26,800	Cuba	4,700
Afghanistan	11,800	Syria	4,000
India	10,000	Vietnam	4,000
Iraq	8,900	Angola	2,900
Iran	6,900	Egypt	2,700

*in current US$million.
Source: Grimmett, op. cit.

Table 6.5: Estimated defence industrial employment

Country	Defence employment in '000s: mid-1980s	Defence employment in '000s: 1990–92
USA	3,100	2,750
Former USSR	6,000	5,900
China	3,000	5,000
United Kingdom	470	400
France	290	255
Germany (East & West)	347	241

Source: H. Wulf, ed., *Arms Industry Limited* (Oxford: SIPRI and Oxford University Press, 1993), p. 14.

Table 6.6: Defence exports and total exports

State	Arms exports share of total exports (est)
Russia/Soviet Union	18.0 – 20%
China	2.0 – 4.0%
United States	2.7 – 4.0%
United Kingdom	1.0 – 2.0%
France	1.0 – 2.0%
West Germany	0.4 – 1.0%

Source: SIPRI and ACDA Data; *IMF Direction of Trade Statistics Yearbook*, International Monetary Fund, Washington, DC, 1992.

Table 6.7: Changes in defence equipment spending in NATO Europe: 1988–92

Country	Change in real defence expenditure 1988–92	Change in equipment expenditure 1988–92	Defence budget (estimate) US$billion 1992	% of defence expenditure on equipment in 1992
Belgium	-23.2	-45.0	2.97	8.6
Denmark	-2.3	12.9	2.59	17.6
France	-0.7	n.a.	31.91	n.a
Germany	-6.1	-54.8	31.03	9.3
Greece	-5.6	-12.9	4.30	21.5
Italy (1991 data)	-8.0	-26.8	19.72	13.8
Luxembourg	7.2	72.3	0.11	4.5
Netherlands	-4.2	-21.2	7.24	16.9
Norway	9.8	37.9	3.51	23.6
Portugal	2.8	-27.3	1.48	6.4
Spain	-9.3	-28.3	9.03	16.6
Turkey	35.8	53.3	5.01	25.4
United Kingdom	4.5	-34.9	41.20	17.6
United States	-7.2	-17.3	305.60*	22.1
Canada	-0.9	-13.2	12.22*	17.6

*1991 figures.
Source: NATO, *Financial and Economic Data Relating to NATO Defence* (Brussels: NATO, 1992). Defence budgets are included only to give a sense of the weighting of different states. With the exception of Turkey, the only countries where equipment spending has risen (Luxembourg and Denmark) are minor defence spenders. The defence budget figures are from the IISS, *The Military Balance, 1992-3* (London: Brassey's, 1992) and are very much early estimates. Final figures will be much affected by the changing values of European currencies against the dollar after September 1992. France does not provide comparable data but it is understood that French procurement spending fell by about 5% a year between 1990 and 1992.

Notes

1 J. M. Rothgeb Jr, *Defining Power: Influence and Force in the Contemporary International System* (New York: St. Martin's Press, 1993), offers a post-Cold War analysis of the utility of military power.

2 See, for instance, Martin van Crefeld, *Technology and War* (New York: Macmillan, 1991).

3 See particularly the SIPRI *Yearbooks*, of which the most recent is SIPRI, *SIPRI Yearbook 1993: World Armaments and Disarmament* (Oxford: Oxford University Press, 1993).

4 Tim Kelsey, 'From Small Wars Come Big Profits', *Independent on Sunday* (magazine section), 6 December 1992, pp. 14–16.

5 See, for instance, Philip Noel-Baker, *The Private Manufacture of Armaments* (London: Gollancz, 1936); and Richard Lewinsohn, *The Profits of War through the Ages* (London: Routledge, 1936).

6 General Strickland quoted in T. Taylor, 'The Supply of Arms to the Middle East and United States' Foreign Policy, 1950–68', London, University of London Ph.D Thesis, April 1972, p. 64.

7 Ibid., p. 72.

8 In addition to specialized works on military assistance and on individual regions, notably the Middle East, some significant works were L. A. Frank, *The Arms Trade in International Relations* (London: Praeger, 1969); Col. J. L. Sutton and G. Kemp, *Arms to Developing Countries, 1945–65* (London, Institute for Strategic Studies, Adelphi Paper No. 28, 1969); George Thayer, *The War Business* (London: Weidenfeld and Nicolson, 1969); John Stanley and Maurice Pearton, *The International Trade in Arms* (London: Chatto and Windus, 1972); and SIPRI, *The Arms Trade with the Third World* (New York: Humanities Press, 1971).

9 A comprehensive modern bibliography on the arms trade would be very extensive, but even in a brief list, in addition to the SIPRI Yearbooks, mention should be made of the following. The most effective comprehensive and scholarly analysis remains Andrew Pierre, *The Global Politics of Arms Sales* (Princeton: Princeton University Press, 1982); Robert Harkavy, *The Arms Trade and International Systems* (Cambridge, MA: Ballinger, 1975) builds on earlier work at MIT by Geoffrey Kemp and Amelia Leiss in analysing the impact of features of the international system on the arms trade. Several works focused on the US export policy and practices, including Michael Klare, *American Arms Supermarket* (Austin: University of Texas Press, 1984); P. L. Ferrari and J. W. Knopf, *US Arms Exports: Policies and Contractors* (Washington, DC: Investor Responsibility Research Center, 1987); and L. Sorley, *Arms Transfers under Nixon: A Policy Analysis* (Lexington: University Press of Kentucky, 1983).

E. Kolodziej, *Making and Marketing Arms* (Princeton: Princeton University Press, 1987), is the definitive book on French policy. Many of these scholarly works are opposed to many of the characteristics of the arms trade, but there are also many books aimed apparently at a wide audience, which may include few footnotes and provide sometimes shocking reading. Many are particularly concerned with the Middle East. Books focusing particularly on the scandalous aspects of arms transfers include David Boulton, *The Lockheed Papers* (London: Jonathan Cape, 1968); R. W. Howe, *Weapons* (London: Abacus, 1981); Anthony Sampson, *The Arms Bazaar* (London: Hodder and Stoughton, 1977 (first edition)); John Sweeney, *Trading with the Enemy* (London: Pan, 1993).

10 See Ian Anthony, 'The United States: Arms Exports and Implications for Arms Production', in H. Wulf, ed., *Arms Industry Limited* (Oxford: SIPRI and Oxford University Press, 1993), pp. 68–9.

11 The latest report is Richard F. Grimmett, *Conventional Arms Transfers to the Third World, 1984–1991* (Washington, DC: Congressional Research Service, Library of Congress, 20 July 1992, Document 92-577 F), henceforth referred to as Grimmett, *Conventional Arms Transfers*.

12 See *SIPRI Yearbook 1992*, Table 10.11, p. 31.

13 Germany's return to the UN Arms Register notes that in 1992 it exported 232 tanks, 355 armoured fighting vehicles, 471 artillery pieces, 20 combat aircraft, two attack helicopters, eight ships and 13,832 missiles (understood to be mainly MLRS rocket artillery delivered to allies in NATO Europe).

14 'Greece Looking to Buy Surplus Saudi MLRS', *Jane's Defence Weekly*, 22 May 1993.

15 Taiwan's imports between 1988 and 1992 according to SIPRI amounted to $2.23 billion.

16 *SIPRI Yearbook 1992*, op. cit., Table 10.11; *CRS Report*, op. cit., pp. 71–2.

17 Mielec and Radom in Poland and Targovishte in Bulgaria are defence industry towns: see 'Arms Makers Feel Chill Wind of Peace', *Independent*, 25 February 1993.

18 Table 10.11 in the forthcoming SIPRI Yearbook.

19 '80% Offsets for F.16 MLU Nations', *Jane's Defence Weekly*, 22 May 1993.

20 'Industrial Offsets May Speed Drive to Build Swedish Tank', *Defense News*, 24–30 May 1993.

21 'Money Problems Hit "Cash-free" Arms Deal', *Jane's Defence Weekly*, 27 February 1993 and 'Cash-free Rescue Bid', *Jane's Defence Weekly*, 6 March 1993.

22 'Russia Chases a World Market', *Jane's Defence Weekly*, 6 March 1993,

gives information on the Russian arms marketing and control arrangements.

23 'Russian Sales Fuel Arms Race', *International Herald Tribune*, 19 October 1993; 'Washington s'inquiète de la vente ....', *Le Monde*, 20 October 1992.

24 'Britain Angry as US Steals Arms Market', *Independent*, 13 December 1992.

25 The surplus on aerospace was $29.6 billion, see 'US Defense Exports Post Surplus in 1992', *Defense News*, 22–28 February 1993.

26 See T. Taylor, *Defence, Technology and International Integration* (London: Pinter, 1981), p. 52.

27 Britain certainly is paying an R&D levy on the Trident missiles which it is buying from the US.

28 Seymour Hersch, *The Samson Option* (London: Faber and Faber, 1991).

29 T. Taylor, 'Arms Supplies', op. cit.

30 Representative Les Aspin, Chairman, House Armed Services Committee, 'Tomorrow's Defense from Today's Industrial Base: Finding the Right Resource Strategy for a New Era', Speech to American Defense Preparedness Association, 12 February, 1992.

31 See 'Slovakia Steps up Arms Production', *Independent,* 7 January 1993; 'Iran Seeks to Join Arms Export Club', *Defense News*, 22–28 February 1993; 'East Nations Find Market Strewn With Sales Hurdles', *Defense News*, 22–28 February 1993; 'Switzerland Sees Arms Exports Fall as New Competitors Enter Market' and 'India Seeks to Double Arms Exports', *Defense News*, 22–28 February 1993; 'S. Africa's Arms Trade Rises', *Defense News*, 23–29 November 1992.

32 'France Snubs China with Sale of Mirage Fighters to Taiwan', *The Times*, 7 January 1993.

33 See Ian Anthony (ed.), *Arms Export Regulations* (Oxford: Oxford University Press, 1991).

34 Noel-Baker, op. cit, especially p. 16.

35 Private interview with a Russian official in 1991.

36 See for instance 'UK MoD Prepares for Massive Surplus Sale', *Jane's Defence Weekly*, 3 April 1993.

37 Seizaburo Sato and Trevor Taylor (eds), *Prospects for Global Order* (London: RIIA, 1993).

38 The procurement elites of the most developed states would argue that, in order for an arms purchaser to be a true 'expert customer', it is necessary for that country to maintain its own substantial research activities near the forefront of defence technological advance so that the purchaser can call on advice from a highly aware scientific community as to the direction and speed of likely future progress.

CHAPTER 7

THE CONTROL OF CONVENTIONAL ARMS PROLIFERATION

TAKESHI ITO

Introduction

According to some policy-makers, the end of the East–West confrontation has saved us from the catastrophe of global nuclear warfare with the concomitant extinction of human life. However, this disaster scenario was never very likely; nuclear strategists considered the chances of global war to be extremely small. But the threat of major confrontation has not receded as much as might be hoped. Many regional conflicts that were dormant during the Cold War era have been reawakened and may prove increasingly threatening.

Since the end of the 1991 Gulf war, the non-proliferation of weapons of mass destruction has become the major focus of global arms control fora. However, the proliferation of conventional arms also poses a significant threat to the world, because such weapons are, in a sense, more dangerous. Countries and political movements are often ready to order the use of conventional weapons, while even the threat of the use of nuclear weapons is extremely rare. During the Gulf war, even Saddam Hussein armed his Scud missiles with conventional warheads, not chemical or biological weapons. The control of conventional arms proliferation is an old issue, but one of increasing importance in our post-Cold War world.

This chapter will consider recent developments in conventional arms control, and present the future scope of discussion.

Recent developments in conventional arms-control initiatives

The UN Register of Conventional Arms

In December 1991, at its 46th session, the General Assembly of the United Nations adopted Resolution 46/36 L entitled 'Transparency in Armaments' by a unanimous vote of 150 in favour. It thus agreed to establish for the first time a UN Register of Conventional Arms.

This landmark resolution has a history of more than 25 years. The issue of arms transfers has been brought up three times at the United Nations, first in 1965 when Malta proposed to the General Assembly that the matter should be addressed by the Eighteen Nations Disarmament Committee. The resolution was not adopted. The issue of arms transfers was raised again in 1968 and in 1976 but in the epoch of the Cold War no agreement could be reached.[1] However, after the Cold War, in 1991, the resolution, initiated by the EC countries and Japan, received worldwide support.

The purpose of the register is to establish a universal and non-discriminatory repository of data on international arms transfers, as well as information on relevant policies, military holdings, and arms procurement through the national production of member states.[2] The requested data, described in the annex to the resolution, cover the following major categories of weapons: tanks, armoured combat vehicles, large-calibre artillery systems, combat aircraft, attack helicopters, warships, and missiles or missile systems. Opening such information to public scrutiny is expected to enhance transparency and reduce misperceptions of other countries' intentions.

Although the resolution was adopted unanimously, there were two abstentions, Cuba and Iraq, and one of the key suppliers of arms to the world, the People's Republic of China, did not participate in the vote itself. The closing date for the first submission of national reports was 30 April 1993. By the end of November nearly 90 states, including Cuba and China, had responded. In fact there were reports from all the main exporting states and most of the leading importers, covering about 90% of world trade in the weapons of concern.[3] Since the register requests member states to provide data on exports of arms as well as imports, positive registration of export information by cooperative countries should allow certain unclear aspects of the flow of arms to be revealed by comparing imports and exports. Considering the fact that the registration of information is on a voluntary basis, further efforts may be required to encourage registration by UN member states, especially from the countries of the former Soviet Union.

The following points should also be raised as future problems of the register[4]:

(1) Member states are only 'invited', not 'called upon', to provide the background information regarding military holdings and procure-

ment through national production. Thus the register requires more information from those who buy their military equipment from overseas rather than making it themselves, perhaps using foreign licences.

(2) The value of transfers will not be reported.

(3) Weapons of mass destruction are not included in the categories of weapons to be reported.

(4) The trade in small arms, arms components, subsystems, arms production technologies and dual-use items will not be reported.

(5) Illegal exports and/or supports for illegal troops cannot be covered.

(6) There are no effective measures beyond comparison with other national reports for verification of reported information.

The main value of the register is that it will contribute to greater transparency; it does not aim at the restriction of the arms trade.[5] Though some scepticism about the register still remains, the UN endeavour deserves respect. It is the first step towards a more comprehensive system of cooperative security.[6] Such an approach needs a firm base, which may be possible to achieve in our new post-Cold War era.

The meetings of the permanent members of the UN Security Council

The permanent members of the UN Security Council, the P5, are also the five major arms-exporting countries in the international market, and share an unavoidable responsibility for controlling arms proliferation. The P5 met as a group to discuss the issue of arms export controls for the first time in July 1991 in Paris. At their second meeting in October 1991 in London they agreed to establish 'Guidelines for Conventional Arms Transfers'. These preclude arms transfers likely to prolong or aggravate an existing armed conflict, to increase tension in a region, or to contribute to regional instability. However, the P5 failed to reach further agreement on advance notifications of arms exports at their third meeting in May 1992 in Washington. The dispute over the export of fighter aircraft from the United States and France to Taiwan, and China's subsequent boycott, have halted the meetings.

It must be admitted that the main focus of the P5 talks has been on reinforcing the legitimacy of arms transfers that are viewed as stabilizing and supportive of a country's defence requirements, rather than on real restraints.[7] The P5 countries have large and influential military industries that are suffering from declining national demand in the post-Cold War era. Although most of them are encouraging the restructuring and conversion of their defence industries, it will be a long time before these efforts become effective; meanwhile, they are keen to sell arms for economic as well as strategic reasons. Nevertheless, every effort should be made to get the P5 talks back on track. The initiative of the P5 is imperative in the quest to control arms exports. The fact that the group has met and discussed the issue is a positive step in arms control, but more remains to be done.

In parallel with the P5 efforts, the Group of Seven, the G7, also released a declaration on conventional arms transfers at the end of their July 1991 London Summit, immediately after the first P5 Paris meeting. Meetings such as those of the G7 and even gatherings of the OECD countries can be useful fora in which the developed countries can seek to coordinate their arms trade policies, even though sensitive issues are raised. Although global control on the arms trade, involving recipients as well as suppliers, is not a real possibility at the present time, developed countries should not miss any opportunity, either at the regional or the global level, to discuss the issue.

Issues in controlling arms proliferation

Supply-side or demand-side?

In dealing with proliferation issues, there are two major approaches: a supply-side effort perhaps involving export controls, and demand-side endeavours where confidence-building measures may play a role. One of the two is always at the core of discussions on controlling the proliferation of conventional arms and of weapons of mass destruction. At the P5 meeting in Paris, for example, the United States presented a supply-side approach that reflected its interest in establishing a mechanism for oversight and influence, whereas France adopted a demand-side stance, which appeared in President Mitterrand's remark: 'The goal should be to safeguard or institute a balance of power in each region, and then to bring it down to the lowest level consistent with the right of all states to security.'[8] This difference in emphasis may also indicate a French intention to resist any US plan to take a dominant position on arms transfers control.

In principle, for export controls to be more effective, the following factors, at least, would be helpful, even essential[9]:

(1) Controlled items should be clearly defined and preferably few in number.
(2) As few countries as possible should be proscribed and processes for identifying them need to be agreed on an international basis.
(3) Controlling countries should be few in number and have a dominant share of controlled items in the world market.
(4) Provision for severe punishment of violators should be made.

However, these factors do not really apply to conventional arms or even to much of the technology associated with weapons of mass destruction.

There are many different types of conventional arms, from machine guns to jet fighters, and in many regional conflicts machine guns kill far more people than do jet fighters. However, it is hard to achieve consensus on the types of arms that need to be controlled: even the UN Register, as noted, covers only seven major categories of weapons. Moreover, modern advanced weaponry is an aggregate of various technologies, from low-tech to high-tech, and it is far more difficult to control the transfer of technology than the transfer of arms, largely on account of the abstract nature of technology. However, technology is arguably a common asset for human beings and every country should have the right to use it for development. Also, technology itself is neutral, and only after it has been embodied into a certain object can we judge whether or not it is being used for peaceful purposes.

Should the developed countries have the right to guide Third World countries in their uses of technology? As long as the use of technology depends on the will of the recipient country, our only recourse at the present is to list certain 'dangerous' technologies (although these are especially difficult to define or identify for conventional arms compared with weapons of mass destruction); and to keep a watch on the flow of those technologies in preparation for the imposition of controls, if necessary. Accordingly, major developed countries are implementing rather strict national export controls on sensitive goods and technologies. However, business people may ask a simple question: if the transfer of arms itself is not yet controlled, why should the transfer of broader technology be restricted? Until we can answer this question satisfactorily, the level of control should be kept to a minimum.

Agreement has been reached in the UN Security Council on arms embargoes against Iraq, Libya, Serbia and Montenegro, but what about other countries? The UN Charter guarantees the inherent right of self-defence, endorses security for all states at the lowest possible level of armaments, and implicitly accepts the legitimacy of conventional arms transfers to meet the security and defence needs of member states. Formal organizations such as the UN can make a charge against a country only when it can be proved to have violated rules; suspicion is not enough. Even if a group of developed countries could agree to set controls on suspect countries, these would be condemned by the Third World as a plot by the North to preserve its own prosperity and perpetuate Southern poverty. Besides, most arms-producing countries consider arms transfers to be a legitimate and useful means of promoting a country's strategic foreign policy goals. Therefore, it is difficult to achieve agreement on which countries are to be proscribed.

Moreover, the capacity to produce many types of conventional arms has spread in recent years to many more states of the world. More countries today have the potential to participate in an arms export race (see below). Consequently, as long as demand exists in the international marketplace, there will be no shortage of suppliers. If one state retreats from the world arms export market, another will take its place.

Thus, export controls, a key element in supply-side efforts, will alone be effective in stopping the proliferation of conventional arms. They can only provide a delaying mechanism until more comprehensive solutions are devised and implemented to address the demand side of the proliferation equation.

The ultimate goal of the demand-side approach is to build widespread, international relationships based on mutual trust and in accordance with the wishes of all countries. However, this will be difficult to achieve. Arms races are unavoidable at present and so we need to contain regional conflicts at a minimum level, and then work to settle them as speedily as possible. Thus a balance is needed between the supply-side and the demand-side approach, involving export controls at a necessary minimum level and a concentration of efforts on certain regions to build confidence. The developed states will have to share information in order to detect the build-up of regional conflicts, expose dangerous projects and eliminate the misperception of intentions. Another element of a balanced approach would be to link economic aid to a recipient's activities in building up its military sector. We should not adopt a crude 'denial' approach to Third World countries which would undermine their legitimate right to development and cause further

North–South problems. However difficult, we must make every effort to persuade arms-receiving countries themselves to accept linkage between economic aid and transparency in arms transfers. An effective transparency regime could also prove to be the key to a transition from 'access denial' to 'conditional access' for armaments necessary for self-defence.[10]

Responsibility of suppliers

Government backing over many years has reduced the ability of even Western military enterprises to survive in civil markets and although the supply-side approach has certain limitations, it should not be given up entirely. Major arms suppliers cannot escape their responsibilities for arms proliferation. In this regard, the P5 initiative could be recognized as an important step towards cooperation. However, military industries in four of the P5 countries are suffering from the decrease in domestic arms demand after the Cold War and have large excess capacity. As long as strong economic pressures exist for those countries to export, further international agreement on restraint is unlikely. A real effort to restructure and convert military industries would be a significant factor in the control of the spread of conventional weapons; indeed arms control and conversion can be seen as two sides of the same coin. But restructuring and the conversion process are more difficult when the world economy is stagnant.

The conversion of military industries involves technological and commercial problems. In furthering science and technology, the objectives of the military industries differ greatly from those of civilian industries; while defence firms pursue high technical performance, civilian businesses seek cost performance. In principle and sometimes in practice applied military technologies may be used in the civil sector. Civil industries can benefit from the spin-off of military technologies in the manufacture of civilian products; likewise military industries can gain from the spin-on of civilian technologies. However, because of its lack of cost-consciousness, military industry often faces difficulties in applying the spin-off of its own technologies to produce marketable civilian products. Such fields as defence electronics and software should be relatively easy to convert, but these are special cases and many other defence fields are not at all easy to change. It might have been a straightforward enterprise to beat swords into ploughshares a thousand years ago, but it is not as simple to transform a tank-manufacturing plant into a competitive tractor factory.

Of great concern is the attitude of the countries in the former Soviet bloc, although they are now cooperating in the global effort to halt proliferation. The Russian government has expressed its intention to promote arms exports, while observing rules of restraint, to earn the hard currency that is desperately needed for conversion of its military-industrial complex. Nevertheless, the Russians have been unsuccessful in expanding their arms export markets since the Gulf war, owing to the poor reliability of their high-tech products. Demand has shifted to arms of higher quality and more advanced technology. Russian Foreign Minister Andrei Kozyrev has reportedly proposed to Washington that it facilitate Russian weapons sales to traditional US markets, in order to decrease pressures on Moscow to resume arms sales to 'the wrong people'.[11] So long as the Western countries fail to present viable means for the conversion of the Russian military industry, they may have to choose between orderly and disorderly Russian arms exports. Should we tentatively support Russian arms exports in order to reduce them in the long term?

The destination of surplus arms resulting from a disarmament agreement such as the CFE treaty is also worrisome. Russia and other former Warsaw Pact countries might prefer to sell those arms at bargain-basement prices rather than scrap them at some cost. Napoleon could use captured bronze cannon to make the monument in the Place Vendôme, but what can be made from Soviet weapons?

China is another country of concern. Although current Chinese weapons do not appear to have contributed to a serious offensive capability in Third World arsenals, and the transfer of Chinese weaponry to the Third World in general has not led to a significant military destabilization of any particular region,[12] China's reluctant attitude towards international arms control talks thwarts progress. In addition, its recent imports of Soviet high-tech weapons and technology will lead to the improved performance, and hence greater marketability, of its own arms.

However, Russia and China cannot be singled out for blame when Western countries are also promoting the export of arms. It appears that the US and French decisions to sell fighters to Taiwan led to the Chinese boycott of the P5 talks.

Another issue that must be addressed is the growing arms production capability in the developing world. As mentioned earlier, developing countries are going to acquire the means to make weapons, whether deliberately or otherwise, through the processes of development and modernization of their manufacturing industries. In addition to China, countries such as Brazil, Israel, Egypt, India and North Korea have built up their domestic military industries and participate in arms export markets. Although they may not have the potential to produce state-of-the-art high-tech weapons, their products can be used effectively in regional low-intensity conflicts and the moderate pricing of lower-tech weapons may well be attractive to many Third World countries. Of course, industrialized nations cannot blame their less developed neighbours for selling arms while they themselves continue to produce and export a bewildering variety of weapons and weapons systems to the world. Are there any fair and logical standards or explanations that could regulate the arms production of developing countries, and should we aim to create an environment that would encourage them to make non-destructive products?

The Japanese perspective

Background

To understand Japan's position on arms transfers, it is useful to look briefly at the history of Japanese policy towards proliferation issues.

Japanese export control policy after the Second World War was based initially on achieving cooperation with the Western allies in trade and other economic activities. Such cooperation was absolutely necessary for the recovery and development of the nation's economy. In order to achieve this goal, endorsement of international arrangements such as COCOM was inevitable. Export control was not considered a major element in national diplomacy. Economic factors were accepted as constituting a legitimate reason for Japan to export arms in accordance with its national security policy. The situation changed significantly in 1987 when Toshiba Machinery's violation of COCOM regulations revealed that Japanese technology had the potential significantly to affect international security. Since then, security has become a key consideration in Japanese export control policy.

Japan also took a firm stance on the issue of arms transfers. This is explained by its strong belief in the need to preserve peace – a belief that manifests itself in the country's constitution, which renounces the right to wage war. In 1967 the Japanese government announced the 'Three Principles of Arms Export',[13] which were strengthened by the subsequent 'Unified View of the Government on Arms Exports'[14] in 1976. The Three Principles were originally intended as a restriction on the arms trade, based on the interpretation of Japanese export control law, but later generally came to be recognized as a complete ban. According to these policies, most arms, arms components, and equipment and technology for arms production may not be exported. However, the United States was dissatisfied with a unilateral flow of arms technology to Japan, and, with the Mutual Defense Assistance Agreement (MDA), required the transfer of Japanese military technology between the two countries. In 1983, Japan agreed to the transfer of arms technology to the United States exclusively, within the framework of the MDA.

In recent years, especially since the war in the Gulf, Japan has sought a more positive and constructive role in the political arena of international relations. In the field of arms control, in addition to the Japanese joint proposal of the UN Register in 1991, the government established the Official Development Aid (ODA) Charter, which states that in extending Japanese foreign aid, full attention should be paid to trends in the recipient country's military expenditures, its development and production of weapons of mass destruction, and its import and export of arms. The Charter is an indicator of the Japanese government's intention to take more strategic initiatives in its diplomacy.

The Three Principles and Japan's role

The Three Principles of Arms Export characterize, in various aspects, the Japanese attitude towards the arms proliferation issue.

The Japanese defence industry has, and has had, a relatively high potential to develop and produce conventional arms. However, because of the Three Principles, it cannot export its products and is now fully dependent on domestic procurement by the government. As a result, despite a fairly large domestic procurement, most of the arms-producing companies do not depend heavily on arms sales and have a variety of civilian products. This will work in their favour as far as future conversion is concerned.

Strong pressure for peace is a constant element of Japanese public opinion and this necessitates Japanese restraint on security issues. As a result, most Japanese support the idea that their country, with its pacifist constitution and policies such as the Three Principles, should take the initiative in non-proliferation talks on conventional arms, as well as on weapons of mass destruction. However, this peace-enforcing stance could be challenged by the fact that the country is also a major importer of arms. Although, because of the Three Principles, the Japanese defence industry has been acting as an 'innocent' player in the world arms market, Japan itself has ranked fourth among the world's arms-importing countries.[15] Moreover, some would argue that since Japan is not a member of the international arms exporters club (which is now suffering from a decline in arms sales as a result of the end of East–West confrontation), it should not take initiatives. How should Japan respond to such a negative view? As the plutonium shipping issue shows, complacent expressions of its 'peaceful' intentions on the part of the Japanese government will not suffice. Well-considered and pragmatic discussions, free from the traditional taboo on a security dialogue, are required before Japan can take forceful initiatives on non-proliferation issues.

Under President Clinton, US Secretary of Defense Les Aspin announced a reorganization and scaling back of the Strategic Defense Initiative (SDI), by shifting the emphasis from stationing anti-missile weapons in space to tactical missile defence systems. The outcome of this reorganization has yet to be seen. However, US pressure requiring the transfer of Japanese military-related high technology and its re-export in 'US' products will probably remain strong, or become stronger. As military forces have come to require more advanced and complex technology, research and development has become a heavy burden for military industries in many countries. Under these circumstances, the formulation among allies of co-development programmes for certain weapons systems might be an attractive way to share the costs. If such programmes should occur between the United States and West European countries, should Japan participate as an ally that shares common values, despite the Three Principles? A secret argument about the Three Principles is going on within the government,[16] and so participation in a co-development programme would be very controversial for Japan. However, in June 1993, Japan was shocked by the news that North Korea had successfully test-launched a mid-range missile that had been under development for many years. This may make Japanese participation in a co-development programme on ballistic missile defence much more likely.

Conclusion

What policies should we strive for in the control of conventional arms proliferation? What is the goal? Unlike weapons of mass destruction, the proliferation of conventional arms in the Third World may not be a direct, significant threat to the West, and a more secure world with fewer regional conflicts must be the goal. Yet such a goal is difficult to achieve with only an arms supply-side approach. A more comprehensive approach is needed, one that is consistent with the aim of Third World development. The UN Register is an example of broader thinking. However, we will need some additional mechanisms that will give greater support to the Register: for example, verification measures, intelligence and even sanctions. If Japan is to take and/or share the initiative in such an approach, it should participate in the international discussions on an equal footing with other countries. This may mean that Japan will have to give up some of its 'Three Principles'. At the very least, Japan should participate in co-developing international surveillance systems and defensive military systems, although the word 'defensive' may be difficult to define.

Notes

1 Eiichi Sato, 'The Present Situation of Arms Export and the UN Register of Conventional Arms' (in Japanese), *Kokusai Mondai*, June 1992.

2 Hendrik Wagenmakers, 'The UN Register of Conventional Arms: A New Instrument For Cooperative Security', *Arms Control Today*, April 1993, pp. 16–21.

3 See UN Register of Conventional Arms, Report of the Secretary General, General Assembly Document A/48/344, New York, 1993; E. J. Laurence, S. T. Wezeman and H. Wulf, Arms Watch: *SIPRI Report on the First Year of the UN Register of Conventional Arms* (Oxford: Oxford University Press for SIPRI, 1993); M. Chalmers and O. Greene, *The United Nations Register of Conventional Arms: An Initial Examination of the First Report* (Bradford Arms Register Studies, No. 2, October 1993); Ian Anthony, 'Assessing the UN Register of Conventional Arms', *Survival*, Vol. 35, No. 4, 1993–4, pp. 113–29.

4 *Asahi Shinbun*, 12 December 1991; *SIPRI Yearbook 1992: World Armaments and Disarmament* (Oxford: Oxford University Press), p. 300.

5 *SIPRI Yearbook*, op. cit.

6 Wagenmakers, op. cit.

7 Peter Almquist and Edwin Bacon, 'Arms Exports in a Post-Soviet Market', *Arms Control Today*, July/August 1992.
8 *SIPRI Yearbook*, op. cit., p. 293.
9 This is based on the idea presented by Yuzo Murayama, an associate professor at Kansai University of Foreign Studies, during the study session on export controls at IIGP in December 1992.
10 Wagenmakers, op. cit.
11 'Russia Seeks US Cooperation for Expanding Arms Sales', *Arms Control Today*, April 1993, p. 24.
12 Richard A. Bitzinger, 'Arms to Go: Chinese Arms Sales to the Third World', *International Security*, Vol. 17, No. 2, Fall 1992, p. 107.
13 Three Principles of Arms Export: the export of arms is restricted by the Export Trade Control Order of the Cabinet and is subject to the approval of the Minister of International Trade and Industry. The approval shall not be granted, in principle, if the destination of export is as follows: (a) a communist area; (b) countries to which the export of arms is prohibited by a resolution of the United Nations; (c) countries which are now involved in or may be involved in international conflicts.
14 The Unified View of the Government on Arms Export: in order to avoid escalation of international conflicts, the Japanese government has been carefully dealing with arms exports from the standpoint of a peace-loving nation. The government will not promote arms export and will hereafter pursue the following policy: (a) the export of arms to destinations covered in the Three Principles shall not be permitted; (b) based on the spirit of the constitution and the Foreign Exchange and Foreign Trade Control Law, the export of arms to destinations other than those covered in the Three Principles shall be refrained from; (c) the export of equipment and facilities which are used for the manufacture of arms shall be treated with the same regard as arms exports.
15 *SIPRI Yearbook*, op. cit.
16 *Nihon Keizai Shinbun*, 27 March 1991.

CHAPTER 8

CONVENTIONAL AND NON-CONVENTIONAL WEAPONS: THEIR CHANGING CHARACTERISTICS AND CHALLENGES FOR ARMS CONTROL

YVES BOYER

Introduction

Aimed initially at lessening tension between the United States and the then Soviet Union, the arms control process originated in the Cold War and was thus conducted according to rules and parameters that fitted the objectives of both superpowers at that time. During most of that period, non-conventional weapons, i.e. chemical, biological and, above all, nuclear weapons, raised the spectre of indiscriminate destruction in the event of their use. Since they also upset many parameters for the conduct of military operations, they were made subject to specific arms control negotiations and regimes. This was particularly the case with nuclear weapons when it appeared, at the end of the 1950s, that their unbridled development was threatening the global environment, given the consequences of atmospheric tests and the already delicate balance of terror between Washington and Moscow. Arms control played a central role in the evolution of the relationships between the USSR and the United States,[1] helping to preserve nuclear stability between the two countries. Arms control was thus primarily directed to providing a framework for preventing the nuclear competition between the Soviet Union and the United States from running out of control in technological or financial terms, through one side perhaps acquiring a destabilizing advantage. Biological and chemical weapons remained marginal in the balance of power that then existed between the two countries. Even if both superpowers maintained investments for research and production in biological and chemical weapons and made plans for their actual use, each considered their use against the other as subject to a kind of taboo.[2] Since conventional weapons did not have a major impact upon the nature and evolution of the confrontation between the two superpowers until the late 1980s when the former Soviet Union began to modify its traditional aim of being able quickly to overrun Western Europe in the event of war, this category of weapons was largely left out of the arms control process except in the ill-fated MBFR talks and the notable CFE-I and CFE-IA agreements of 1990–1.

In the post-Cold War era, arms control clearly has many different purposes, among them to contribute, with many other factors, to the defusion of tensions in areas of actual or potential international turmoil and to prevent the use of chemical or biological weapons which are forbidden by international agreement. Arms control has now become part of preventive diplomacy and can thus no longer be limited to a small number of players. Its main focus has accordingly shifted from nuclear issues towards biological, chemical and, particularly, conventional weapons. In this last domain, changes in the substance and focus of arms control will undoubtedly bring new difficulties, since the realm of conventional weapons is itself rapidly evolving as the result of the pace of technological change which deeply affects the nature of warfare, with qualitatively new weapons systems acquiring a paramount influence; and also because of the quasi-nuclear problem of conventional explosives spreading radioactive materials.

Indeed, from the early 1980s, with the shift of the centre of gravity of superpower competition towards science and technology, the pace of technological developments in weaponry and the transformation of warfare have gradually increased the significance of new areas not traditionally covered by arms control. Modern armed forces rely on technologically advanced equipment in command, control and communication for guidance and target acquisition for 'smart' land, air and naval weapons systems, as well as for planning complex military operations. Future military operations will require sophisticated means of surveillance and targeting, and more guided weapons with the increased range associated with complex weapons platforms. Today, developed states deploy high-tech, multi-category and multi-mission weapons such as the cruise missile.[3] This evolution is bringing a revolution in the conduct of military operations, transforming the nature of conventional war. The traditional battlefield is being replaced by a 'cybernetic' battlefield.

This will have profound consequences for the ways in which arms control will be managed and adapted in the future. Since biological and chemical weapons have been drawn to the attention of the community of nations, pressures against their production and their use remain strong, as was shown during the Gulf war: Iraq did not dare to use its chemical weapons against the allied forces despite vehement declarations that it would do so. The taboo against chemical weapons has been formally reinforced with the signature of the Chemical Weapons Convention, a process which has cast light on the position of particular countries such as Iran which have refused to associate themselves with the 135 signatories to the convention.[4] But if a country appears to be contemplating the production and even the use of chemical and biological weapons, its threat will probably be less credible from a military standpoint unless it also possesses a sophisticated and reliable command, control, communication and intelligence system associated with potent means of delivery – areas where the Western world enjoys a considerable edge in terms of countermeasures against any plausible opponent.

At the same time, while the traditional distinctions between categories of weapons (nuclear, chemical, biological and conventional) remain largely valid, they are in some danger of erosion. For instance, technologically poor states or terrorist groups may be tempted to spread poisonous radiation in an area by detonating a conventional explosive surrounded by a toxic material such as plutonium. This would not be a 'nuclear' weapon but could use radiation as an instrument of terror. With the former Soviet Union having perhaps 100 tonnes of surplus plutonium from redundant weapons, and nuclear power stations around the world generating plutonium as a by-product of power production, the possibility that plutonium might be seized and used for terror purposes cannot be discounted. Another possibility which straddles the conventional–nuclear divide concerns the fallout consequences of a conventional attack on a nuclear power station. In the chemical and biological weapons area, genetic engineering and other advances in biochemistry could seriously erode the chemical–biological weapons distinction.

With changing elements defining weapons performance (range, yield, accuracy), the distinction between 'strategic' and 'tactical' weapons systems has been weakened. In certain cases the differences in the effects of conventional and some nuclear weapons have been reduced significantly.[5] At a time when complex systems are transforming the nature of warfare and force multipliers are acquiring a key role in the design of modern force structures, conventional arms control will become an irrelevant tool for promoting peace and stability if it does not extend its scope and purpose. This presupposes an understanding of the new qualitative changes in conventional weapons systems as well as their military utility on the 'extended battlefield'.

The transformation of conventional weapons and warfare

Since the mid-1970s, developments in advanced technologies for conventional weapons, particularly in four areas (communications, reconnaissance, guidance systems, command and control of operations), have introduced radical changes in the conduct of war by transforming weapons into weapons systems. The systems combine several different functions such as the transmission and processing of information, target acquisition, fire control and terminal guidance. The weapons system has become a sub-component of a modern force structure which seeks to generate a synergistic effect among complex command and control centres, reconnaissance and strike systems, and army, air and naval units. As pointed out by analysts drawing lessons from the Gulf war: 'ultimately, the characteristics of the individual high-technology weapons are less important than those of the elaborate electronic and intelligence network responsible for the effective use of their "intelligence"'.[6]

These changes have been made possible by revolutionary progress in micro-electronics with an exponential increase in data-processing capabilities and progress in software. Miniaturization and data processing have stemmed from the civilian and military research programmes of the 1980s, such as the US Very High Speed Integrated Circuits (VHSIC) programme. At the time of its inception, the VHSIC programme was conceived as a way to widen the technological gap between the United States and the then USSR: it was a factor behind the change in the US export restriction policy with the hardening of the COCOM regulations. This was strategically important since electronics now accounts for half of the equipment costs in Western armed forces.

The most difficult task in moving from traditional conventional weapons to high-tech weapons is to design and build the electronic components for dealing adequately with each sequence necessary for the overall success of the weapon, and also the connections which enable each sub-component to communicate with the others. This communication makes modern weapons systems 'smart', allowing Precision Guided Munitions (PGMs) to be particularly accurate with a high chance of success at the first shot.

The development of a huge array of sophisticated electronic means of surveillance is also altering the notion of what constitutes the traditional battlefield. Having a front and a rear, the modern battlefield is 'extended', with no part safe from precise and deadly attacks. Accordingly, the battlefield is divided into zones, each of which is the object of specific means of surveillance and subject to the effects of specific weapons systems. This 'extension' of the battlefield requires both the ability to detect an enemy's activities deep behind the line of immediate contact and sophisticated electronic means to analyse, process and efficiently combine the collected data. The use of high-tech weapons is thus part of the decision-making process involving the reception, processing and analysis of reconnaissance and intelligence data in management centres, from which are given orders on troop movements and weapons employment. Headquarters have to rely on complex electronic devices such as the US Joint Tactical Fusion Centre (JFT) or the French SACRA system[7] which can process huge quantities of information. The information provides military commanders, almost in real-time, with enemy locations and force deployments in the theatre from the Forward Edge of the Battle Area (FEBA) to very deep into hostile territory (more than 100 kilometres behind the FEBA). Targets whose destruction will influence the course of the battle and force movements can be selected very rapidly. This was, again, illustrated during the war against Iraq which lasted in its aerial phase as long as needed to dismantle the Iraqi military machine as a coherent force. When the enemy order of battle was sufficiently damaged, the land phase of the war was opened. It lasted 100 hours, one-tenth of the air phase. The conduct of sophisticated military operations is already managed with almost real-time processed information, giving war a very high tempo, now labelled by US Air Force planners as 'hyperwar'.[8]

Henceforth, on the battlefield, there should be a clear advantage in firing the first salvo and a target acquired should be a target destroyed. This consideration may have a destabilizing effect during an international crisis since the certainty of the effect of a weapon and particularly the confidence that collateral damage will remain very limited may precipitate a decision to go to war. Saturation bombing (with indiscriminate collateral damage) can now be replaced by the use of pinpoint-accuracy weapons systems. In the early 1980s studies demonstrated that the destruction of a Warsaw Pact army group threatening NATO's front in central Europe would have required 33,000 tonnes of conventional munitions and 5,500 air sorties, but only 500 tonnes of terminally guided bombs and about 100 air sorties.[9] For artillery shells, it is estimated that when 380 artillery shells from 155 mm guns are required to destroy a specific target, fifteen shells with directed effects and only eight terminally guided projectiles are necessary.[10] What was pure speculation in the early 1980s became reality in the early 1990s, as was shown during the war against Iraq where a decisive success in the conflict resulted from the massive use of PGMs, including nearly 300 conventionally armed Tomahawk cruise missiles; about 4,000 air-to-ground laser or TV guided bombs or missiles (such as AS. 30L, and Have Nap); more than 1,000 anti-radiation air-to-ground missiles (such as Alarm and Harm); more than 5,000 Maverick air-to-surface missiles; and more than 8,000 anti-tank missiles (including Hellfire, TOW, HOT and Milan).[11]

Advanced conventional weapons now provide the means for launching very destructive strikes in the depth of a theatre of operations. Their great accuracy and effectiveness are increasingly negating the military necessity of employing tactical nuclear weapons even against hardened targets. Undoubtedly, in certain circumstances, high-tech conventional weapons will erode the technical relevance of nuclear weapons for strikes at strategic targets. As pointed out by Carl Builder[12]:

> tactical weapons are demonstrating techniques for both controlled dispersal and precise delivery of some extremely efficient conventional munitions for penetration, fragmentation and concussion; space systems are providing the global information necessary for precision navigation and high-capacity communication, reconnaissance and surveillance; strategic delivery vehicles have proved their ability to transport substantial payloads rapidly and reliably over intercontinental distances.

In the near future it seems that countries with advanced technology will aim to design and operate cruise missile weapons systems, like the 'Super Apache' developed by Matra-Défence and its partners. This system will combine a stealth-strike capacity (whose utility was demonstrated by the decisive role of the F117A aircraft in the early stages of the war against Iraq) with autonomous navigation capabilities and an ability in the terminal flight phase to combine two modes of guidance (infra-red homing head and active millimetric radar). This will allow conventional strikes at targets of strategic importance with a precision of about one metre after a 600-kilometre flight.[13] Such capabilities may, in the future, complement the nuclear strategy of those countries with a nuclear deterrent force.[14]

Precision-guided smart munitions carried by missiles or aeroplanes are not the only means of challenging the efficiency of tactical nuclear weapons for some specific battlefield uses. In a very limited zone, new types of explosive can produce the same effect as small nuclear weapons in terms of heat, electromagnetic impulse (EMP) and overpressure. The obvious example of this type of weapon is the US Fuel-Air Explosive (FAE) BLU-82 air-dropped bomb. The BLU-82 weighs 6,800 kg of which 5,700 is the GSX explosive. When exploding, the BLU-82 produces an overpressure of 70 kg/cm^2 by absorbing the oxygen in the air, sends a shock wave circulating at a speed of 1,800 metres per second and damages radio-electronic systems.[15]

The transformation in the nature of conventional warfare with the revolutionary effects of electronics and information processing is also having an impact on traditional military equipment. If one takes the main battle tank (MBT), its three principal characteristics (firepower, protection and mobility) have been significantly improved by new technologies. There has been a noticeable increase in firepower as the projectile launched by the gun has become more efficient, and with the computerization of fire control systems. Significant progress in metallurgy means that the traditional shaped charge projectile is being replaced by a new warhead, the Armour Piercing Fin Stabilized Discarding Sabot (APFSDS), which is made of tungsten alloy or depleted uranium, to devastating effect. A direct hit with the first shot on the move is possible with new complex fire control systems using computers which combine data from a laser rangefinder, an image intensifier and a thermal camera. The automatic loader helps with the selection of the appropriate projectile. Such equipment increases the proportion of electronics in the cost of the tank. Indeed, electronics accounts for 50 per cent of the cost of the new generation of tanks such as the French Leclerc. These vehicles also increasingly use stealth features to lessen their thermal or infra-red signature. They also deploy new devices warning of detection and triggering active countermeasures such as the VIRSS (Visual and Infra-Red Screening Smoke). Just like modern fighter aircraft, modern tanks may have jammers with lasers and chaff.

The blurred distinction between conventional and non-conventional systems and its possible impact on arms control

The increasingly lethal nature of modern conventional armaments, the improvement in surveillance and data processing, and the limited number of sophisticated weapons systems in the field due to their high costs will make the future battlefield less dense, more deadly and more fluid. Armed forces will be ever more tightly integrated within a complex system of command through which operations will be jointly executed by combined forces from all services (air, sea and land forces). This joint action will make sophisticated forces like a movable fortress:

> composed of ever-more complex and mutually supporting offensive and defensive systems ... [this force] will construct and operate from a base that the enemy cannot decisively penetrate, either physically, visually or electronically. From that base, we can strike swiftly, accurately and decisively, expanding much of the fortress protection to cover our air and ground manoeuvre forces as they move. Wherever we must halt, we will build our defence again in four dimensions – vertical, horizontal, in depth and in the electromagnetic spectrum.[16]

The battlefield will be extended throughout the whole theatre of operations, as foreshadowed by the war against Iraq. Such characteristics will demand skilled commanders and will lead to a reshuffle of current military organizations promoting joint operations above the tactical echelon on a scale never known before. The objective is no longer to attack the bulk of enemy forces themselves, as was traditionally the case with the doctrine of attrition, but to degrade and destroy their cohesion and their 'nervous systems', i.e. their command and control structures, cutting them off from intelligence and coherent orders. Friendly forces will seek to move and strike more rapidly than their adversary, whose movements have to be anticipated since the size of a unit – a division or an army corps – determines its pace and the time it requires to prepare for action at a specific level of operation, whether tactical or operational.[17] Initiative is, more than ever, the keyword on the battlefield and the need is for a command structure which determines objectives (*Auftragstaktik*) rather than defining strict procedures of actions emanating from the commander-in-chief (*Befehlstaktik*). The structures and equipment of a modern army are thus increasingly relying on force multipliers involving sophisticated technologies which cannot easily be brought into an arms control regime.

Going further, revolutionary technological advances may lead to a total reappraisal of the criteria on which military power is grounded. In this context an area particularly worth attention is that of directed energy weapons. Directed energy weapons fall into three broad categories: lasers, radio-frequency/microwave energy weapons (generically called High-Powered Microwave, HPM), and charged particle beam weapons. Should they become operational, their main functions, according to a document from the US Joint Chiefs of Staff, will probably be to disrupt, upset and destroy enemy systems.[18] Those weapons rely largely on non-lethal or 'disabling technologies'. McDonnell Douglas, for example, is currently working on a research programme code-named COBRA designed to blind electronic and optical systems used by hostile helicopters, tanks, missiles or artillery for detecting and tracking.[19] Such types of equipment are already on board some Royal Navy ships. While these types of weapons systems are still largely in their infancy, the US Army Training and Doctrine Command (TRADOC) is circulating a draft study ('Operations Concept for Disabling Measures') on their impact on the battlefield of the future. Research is also being undertaken on 'soft-killing' mechanisms, i.e. electronic viruses or codes placed covertly in software and then activated in order to disable high-tech weapons systems. This would produce devastating effects on much sophisticated equipment such as modern aircraft whose aerodynamics are unstable and which can be flown only 'by wire' through 'complex computer' systems. Should such systems be disrupted by 'soft-killing', the aircraft would be lost.

These developments will have a significant bearing on arms control in the future. Nuclear deterrence is no longer a central factor in the realm of strategic and military affairs, and if the two biggest nuclear powers move further towards a posture of minimal deterrence, as Britain and France already have, nuclear arms control will become only marginally significant.

But some conventional weapons can cause effects comparable to small nuclear weapons and the rapid transformation of conventional equipment and warfare will require innovative measures to constrain the development and deployment of certain types of sophisticated equipment which could have a destabilizing effect on the balance of forces in a region. With qualitative improvement becoming more significant than quantitative development, conventional arms control needs new emphases and approaches. Traditional methods of arms control cannot comfortably address new military systems which can be easily controlled, such as lasers, or those whose impact does not depend on the number deployed but instead on their quality as command and control systems. As some analysts[20] have suggested, it may be possible to meet this challenge by concentrating on the missions being performed rather than the weapons systems themselves.

Notes

1 On this subject, see the now classic book: D. G. Brennan, ed., *Arms Control, Disarmament, and National Security* (New York: George Braziller, 1961), p. 475.

2 The Biological Weapons Convention (BWC) prohibited the development, production and stockpiling of bacteriological and toxin weapons. It was opened for signature in April 1972 and entered into force on 26 March 1975. The BWC provided a useful definition of biological weapons as 'microbial or other biological agents', i.e. referring 'to living organisms or infected materials that rely for their effect on multiplying inside the person, animal or plant attacked' (*The Arms Control Reporter*, (c) idds 1–93, January 1993, p. 701.A1.)

3 See Rose E. Gottemoeller, 'Land-Attack Cruise Missiles', *Adelphi Papers*, No. 226 (London: International Institute for Strategic Studies, Winter 1987–8), p. 56.

4 The title of the convention signed in Paris on 13 January 1993 is: 'Convention on the Prohibition of the Development, Production, Stockpiling and Use of Chemical Weapons and on their Destruction'.

5 See in particular the premonitory analysis made by Christoph Bertram, 'Arms Control and Technological Change: Elements of a New Approach', *Adelphi Papers*, No. 146 (London: International Institute for Strategic Studies, 1978), and Carl H. Builder, 'The Prospects and Implications of Non-Nuclear Means for Strategic Conflict', *Adelphi Papers*, No. 200 (London: International Institute for Strategic Studies, 1985).

6 Gene I. Rochlin and Chris C. Demchak, 'The Gulf War: Technological and Organizational Implications', *Survival*, Vol. XXXIII, No. 3, May/June 1991.

7 SACRA (Système d'Aide au Commandement et au Renseignement de l'Avant) is designed to facilitate the decision-making of army units at the corps level.

8 Eric H. Arnett, 'Welcome to Hyperwar', *The Bulletin of the Atomic Scientists*, September 1992.

9 Benjamin Schemmer, 'Defense Resources Board, Congress, Order Early IOC for Assault Breaker', *Armed Forces Journal International*, September 1982.

10 Pierre-André Moreau, 'Feux dans la profondeur et munitions intelligentes', *L'Armement*, May–June 1992.

11 Philippe Turpin, 'Guerre du Golfe: la révélation des armes intelligentes', *L'Armement*, July–August 1991.

12 Builder, op. cit.

13 Pierre Langereux, 'Le Super Apache de MATRA frappera à 600 km avec une précision de 1 mètre', *Air et Cosmos*, 7–13 June 1993.
14 See for example Jean-Louis Promé, 'Towards a French Non-Nuclear Deterrent?', *Military Technology*, June 1993.
15 'L'arme la plus meurtrière de l'arsenal américain: la bombe à aérosol BLU-82', *Air et Cosmos*, 4–10 March 1991.
16 Major Ralph Peters, 'The Movable Fortress: Warfare in the 21st Century', *Military Review*, June 1993.
17 Operations FM 100-5, Department of the Army, 20 August 1982. (Cf. the notions of 'areas of influence' and 'areas of interest' in the US Army Airland Battle doctrine.)
18 James W. Rawles, 'Directed Energy Weapons: Battlefield Beams', *Defense Economics*, August 1989.
19 Mark Tapscott and Kay Atwal, 'New Weapons that Win Without Killing on DOD's Horizon', *Defense Electronics*, February 1993.
20 See Bertram, op. cit.

CHAPTER 9

FROM POLICIES TO SPECIFIC DECISIONS : PRACTICE AND PROBLEMS

CHARLES DRACE-FRANCIS

Strategy, economics and morality

It is impossible to discuss the practice and problems of export controls without reference to some of the determining factors which have been invoked by those involved and by commentators. This chapter therefore starts with a brief discussion of these factors and their historical development as seen by an exporter of conventional armaments. Historical reference is needed since much of UK law is based deep in the past and greatly influenced by precedent.

Three considerations have played a notable role in the exercise of export controls, especially in the UK:

- strategy;
- morality;
- economic self-interest.

The strategy consideration has a very long history. There are examples of prohibitions on the sale of military equipment from England to Scotland in the fourteenth century. The longbow was in its day a stand-off weapon of great potency, comparable in effect to the ballistic missile of the twentieth century, particularly in its deterrent, defensive role. But the problem was mainly one of import, not export, given the shortage of indigenous material to make longbows. A statute of 1483 required four bowstaves to be imported by aliens with every ton of goods and ten bowstaves with every butt of wine.[1] English craftsmen were even sent to Poland to make bows and send them home.[2]

Similar denial and acquisition strategies were invoked during the development of the long-range cannon. But the favourite method of denial control was not necessarily a prohibition on exports of the technologically significant goods, e.g. cannon themselves, which would have been more or less impossible given the tenuous control by central authorities over local conditions of life. Instead there were prohibitions relating to the movement of experts, who knew how to make the cannon and armour. In other words, the control of mercenaries was more important than the control of material. This is a point which has not been very much considered in recent years, when the greater degree of central authority and the greater clarity and definition of international borders have enabled states to make it more difficult to move material without permission, although intangible technology transfer is covered by US and German legislation. Indeed, in the UK, the movement of persons has been relatively neglected since 1714 when emigration was deregulated (although there was a tightening up in the 1860s as a result of the Alabama case[3]). But it must be remembered that some of the greatest developments in military technology proliferation were caused by the diaspora of German scientists after the defeat of the Nazi regime in 1945.

The morality consideration also came into evidence at an early stage in English history, particularly during the Reformation, when England's brand of Christianity began to differ from that of its neighbours and those with whom it was most likely to go to war, e.g. the Scots, Irish, French and Spaniards. By contrast technology passed relatively freely from the Muslims to the Christians during the Crusades, despite the fact that for strategic reasons it would have been more advisable to deny the opposing religion advanced weapons technology.

Economic interest arguments in relation to arms exports developed with the growth of the science of economics during the twentieth century. The original strategic ideas promoted in particular by Colbert in seventeenth-century France favoured the notion that the state should foster craftsmen in all important skills, including weapons production. But with the growth of international trade, self-sufficiency ceased to be the economic be-all and end-all. Whole communities became dependent on arms exports, and the argument developed that the economic health of these communities would be damaged by excessive controls.[4] In other words, whereas strategic principles were more or less neutral from a moral point of view, economic principles appeared always to be opposed to moral ones. What was difficult to find was a system of control that met the requirements of morality while at the same time safeguarding the economic interests involved.[5]

A fairly representative example of these three arguments tending in different directions was the proposed sale of frigates to the Spanish government by the UK in 1965. At the time, Spain was under the Franco regime. The new Labour government, which had come into power in Britain in October 1964, was strongly opposed to the Franco regime on moral grounds, many of its older supporters having condemned the Franco uprising. But the negotiations for the sale of the frigates had been entered into before the change of British government and many jobs were dependent upon the contract at a time when the advent of the socialist government had led to a lack of economic confidence. There was also a strategic consideration involved in that, with the growth of the Soviet navy, it was necessary for the Atlantic Ocean adjacent to Spain to be patrolled, even though Spain was not at the time a member of NATO. Nevertheless, the ultimate decision by the British government was to cancel the order, which the Spanish then said they did not need anyway.

This did not have a great material effect on British economic interests in Spain, but cancellations can be economically relevant in themselves. Any consideration of economic interest must also therefore take into account the effects that would flow from licensing or not licensing a particular order. For instance, the decision of the French government to sell Mirage air combat aircraft to Taiwan has affected French economic interests in China although it will be interesting to see how much. It is often these aspects of economic interest which lead to fierce debate between different agencies inside individual governments. For instance, in early 1993 the Boeing aircraft company and General Electric applied for a licence to sell Boeing 737 civil airliners with GE engines to Iran. There was a traditional line-up of American government agencies for and against the granting of a licence. The Department of Defense and the National Security Agency argued against allowing the Iranians access to modern aviation technology, for strategic reasons. The State Department also argued against the sale for political reasons, in order to enforce the isolation of a terrorist state. But US economic agencies tended to support the economic interests of the companies concerned. Suffice it to say, definitions of economic interest are often contradictory and inconsistent. Inside the COCOM regime, for instance, the case of the Soviet Tyumen pipeline in 1982–3 underlined the point that what might be good for the UK engineering industry and in particular the John Brown company, which had contracted to supply the turbines for the pipeline, might also be bad for the interests of American rivals in this engineering field, notably Bechtel. Since what was at stake was the largest pipeline project in the world, naturally those who were prevented from bidding for it for reasons of control of dual-use technology claimed that their economic interest was being harmed.

It may seem obvious that the strategic factor should always prevail. But often the greater the prohibition and the greater the risk for those who wish to get round it, the higher is the profit to be made by doing so successfully. It is the lure of very large profits which makes export controls difficult to administer and increases the relative importance of economic and moral factors. It is important, when discussing the economic aspects of export controls, to understand that they are a barrier to free trade and therefore, in themselves, distort the laws of economics, in particular as regards comparative advantage and the distribution of labour. What this means is that actual or potential controls can make it appealing to governments to produce arms within their countries and therefore to proliferate arms industries, rather than allowing a natural globalization of the trade. But paradoxically it was the Soviet Union, which paid the least attention to the economics of arms production and had almost no concept of 'price', which was the largest arms supplier throughout the 1970s and 1980s.[6]

To some extent the laws of economics cannot be suspended even in controlled goods. The value of arms is affected by demand, as well as by supply. Following the Iraqi attempt to overrun Kuwait in 1962, the British government left a large amount of equipment, particularly tanks, in Kuwait, because it was not considered economic to ship them back to the UK. Also the UK Treasury was keen to cut the costs of the original defence operation by selling the tanks to the Kuwaiti government, which at that time was still under the protection of the British government. The Emir of Kuwait saw no reason to purchase the tanks after the Iraqi threat had receded. But one night when there was a rumour that the Iraqis were returning, he went to the house of the British political representative in Kuwait and asked to buy the tanks there and then for £2 million. The UK Treasury had originally been holding out for only £700,000, but the emergency greatly enhanced the value of the tanks. A deal was drawn up and signed at 2 a.m. by the Emir himself.[7] The need to get rid of surplus equipment and at the same time enjoy foreign exchange advantages is now an important factor in the former Soviet Union.

Returning to the strategy concept, it is important to note that this should not just be considered a negative item based on denying technology to perceived or potential enemies. Friends and allies need to be adequately armed. For instance, had the Kuwaitis been sufficiently well armed in 1990, that is, had they had the means to deliver nuclear weapons, the Iraqi invading forces would probably not have attacked.[8] This argument is analogous to the UN Article on self-defence, but goes wider. It is, for instance, in the Western interest that Saudi Arabia, which has the most readily available supply of cheap energy, should be adequately protected from foreign aggression. In addition, because the supply of arms confers influence, if you allow a rival to supply arms, you may lose political leverage in the recipient country. This argument was much in evidence during the arming of the Shah's regime in Iran. In other words, there are strategic reasons for promoting exports as well as for denying them. Finally there is the well-known argument which one might consider the last refuge of the arms dealer, namely that if he or she does not supply the arms, someone else will.

The UK export control machine

The practice of export controls cannot be understood by reference to principle alone. The rest of this chapter focuses on the way in which controls are operated in the UK. The United Kingdom was the first country to impose specific export controls on conventional weapons in modern times.[9] Many of its current controls stem from the Import, Export and Customs Powers (Defence) Act, 1939, one of a number of pieces of emergency legislation passed at the outbreak of the Second World War – indeed the only piece of that legislation which has remained on the statute book. Goods which are subject to control are listed in the Export of Goods (Control) Order (EG(C)O). This order was most recently revised at the end of 1992. Officially, the purposes of export controls include meeting the needs of:

(1) the collective security of the United Kingdom and its allies in NATO;
(2) national security;
(3) foreign policy requirements;
(4) international treaty obligations and commitments;
(5) the United Kingdom's non-proliferation policy;
(6) concerns about terrorism or internal repression;
(7) national heritage.[10]

The EC(C)O is administered by the Export Control Organization (ECO) in the Department of Trade and Industry (DTI). This organization was set up in 1988. It now has about 130 generalist civil servants, up from about 50 in 1990. One of the considerations which has come to the fore with the increase in licensing as a result of the Gulf war and the war in the former Yugoslavia is the need to increase the number of civil servants processing licence applications. This is an economic cost which is important in government considerations, though not necessarily as important as the economic cost involved in the loss of export opportunities themselves.

Currently the ECO is divided into eight units:

(1) the Policy Unit is responsible for developing export control strategy and ensuring the adequacy of controls. Its concerns include non-proliferation issues and the future of controls in the EC single market after 1992;
(2) the Export Licensing Unit (ELU) processes export licence applications, seeking advice as necessary from other ECO units or from other government departments. It comprises two sections. One deals with all applications in respect of industrial dual-use goods, and with COCOM proscribed destinations and other destinations regarded as sensitive. The other section deals with military goods and with atomic energy goods for non-sensitive destinations;
(3) the Sensitive Technologies Unit comprises a team of technical specialists – scientists, engineers, etc. – who provide advice on whether equipment or technology falls within the statutory restrictions published in Schedule 1 of the EG(C)O. This advice is available to ECO staff and to customs officers at ports and airports, as well as to industry;
(4) the Compliance Unit monitors compliance by exporters with the conditions of export licences and generally ensures that exporters remain aware of their responsibilities with regard to export controls;
(5) the Sanctions and Embargo Enforcement Unit deals with measures introduced under UN Security Council Resolution 661. It is also responsible for monitoring cases where violation of controls is suspected. In such cases, information is passed to customs who, as well as carrying out the physical controls at the UK frontier or at inland offices, especially as regards intra-EC trade from 1 January 1993 (see below), are empowered to investigate suspected offences;

(6) the Computer Unit provides support for all IT functions of the ECO;
(7) the Scott Inquiry Unit (see below);
(8) the Coordination Unit, which provides management information and other common services to the whole of the ECO.

In line with recent COCOM amendments to the Industrial List of controlled items (the 'Core' List), which have led to a substantial reduction in their number, the ECO is making greater use of Open General Export Licences (OGELs) for certain goods, and has also introduced Open Individual Export Licences (OIELs) for dual-use industrial goods and for military goods as part of more flexible licensing arrangements to facilitate the licensing of exports to non-sensitive destinations. Broadly, these licences remove the need to apply for specific/individual (quantitative) export licences, thereby reducing the burden on the exporter and on the ECO.

OGELs allow the export of eligible goods without the need to seek prior approval from the DTI, provided the conditions set out in the relevant Open General Licence are met. The terms of the licences are published by, and are available from, the DTI and exports of eligible goods can be made without further reference to the DTI by any UK exporter. Exporters must, however, declare the liability of goods to export-licensing control and use of the OGEL on the Customs Export Entry document for consignments to non-EC destinations. Customs Entry for goods dispatched to EC destinations is no longer required (since 1 January 1993). Export licensing controls remain in place, however, even in the absence of routine customs checks at UK and EC frontiers, and exporters must now comply with the licensing control arrangements required by the local customs office for the premises from which licensable goods are dispatched. The grant of an OIEL will be considered by the ECO, subject to prior application by the exporter, where consignments of specified goods are regularly exported to specified destinations. The duration of the licence is three years and while it is unlimited as to quantity/value in that period, a number of conditions attach to its use. If the application details fit the criteria demanded by the ECO, an individual licence is issued for retention by the applicant. Customs headquarters is notified of issue with details of the licensed goods and destinations, and in turn advises its export control officers at UK ports/airports to permit export without actual production of the licence at the point of export. This is subject to declaration of the OIEL number on the Customs Entry (non-EC destinations) or on the Commercial Documentation (EC destinations).

The ECO works with other UK government departments (and some foreign governments) to prevent the export of goods to destinations for purposes which are not consistent with UK interests. It seeks to ensure that no goods are wrongly licensed, that exporters are correctly advised as to the need for licences and that exports are not frustrated by the improper exercise of licensing controls.

Broadly, all commercial exports of scheduled (controlled) goods by UK traders are the subject of licensing control. However, where exports of controlled goods are made by traders on behalf of, or to the order of, UK government departments, the ECO considers that it would not be appropriate to intervene or impose licensing controls on these arrangements. Accordingly, such exports are considered to be outside the scope of export licensing control. On some occasions, 'administrative waivers' of the need for export licences are advised by letter to the exporter concerned, for instance in the case of dispatches of strategic goods pursuant to international collaborative defence projects of which the British government is a member, such as Tornado/Eurofighter 2000. Also goods may be exported on behalf of the Ministry of Defence pursuant to a contract between the British government and a foreign government for the supply of strategic goods, such as the Al Yamama contract with Saudi Arabia.

Other legal and administrative exceptions to the EG(C)O include: authorized ships/aircraft stores; ships/aircraft/hovercraft engaged on normal commercial journeys; goods consigned to British ambassadors and certain other officials; goods remaining on board for exportation; goods (other than aircraft) exported after repair to the country from which they were imported (other than Iran, Iraq, South Africa and Namibia); and goods consigned to workpoints in the UK-designated area of the continental shelf.

Typed applications (to facilitate optical character recognition) on an officially provided form (two parts, six pages) plus up to five photocopies are required. Applicants must give details of the nature of their business, the nature of the consignees' and end-users' business, whether there is any 'association' between them, and whether the goods are suitable for/capable of military, chemical, biological or nuclear use, or whether it is known/suspected that they might be used for such purposes.

Applications for the export of military atomic energy and/or dual-use industrial goods and related technology must normally be supported by a detailed technical specification of the goods, together with undertakings or end-user statements to the effect that the goods are required for a specific use and will not be re-exported. If the application concerns firearms, a copy of the Registered Firearms Dealers Certificate is required for 'Section 1' firearms. A copy of the Home Office Authority is required if the application concerns 'Section 5' firearms. For exports to the 'proscribed destinations' additional supporting documentation concerning end-use, including disposal control, is required.

All questions on the application form have to be answered: false statements in applications carry penalties, as does failure to comply with any condition attaching to the licence. In the early stages of an arms export negotiation with any potential overseas customer, close liaison is maintained with the Defence Export Services Organization (DESO) of the UK Ministry of Defence (MOD) to ensure that approval of any sale is likely to be forthcoming. A necessary preliminary action taken by British Aerospace (BAe) is to seek and obtain MOD permission for the release of technical information relating to performance and use of specific military equipment to nominated potential customers. Before any technical information not in the public domain is released, formal authority from the MOD (through an MOD Form 680 reply) is obtained. Subject to the continuing approval of DESO, serious negotiation for the sale of military equipment to meet the customer's requirements can follow.

When a firm contract is in place, and delivery schedules established, formal action to secure the necessary export licence (or waiver) from the DTI is taken. Evidence of DESO involvement is furnished with the licence application, together with any supporting documentation that may be necessary, such as copy extracts from contracts, end-user statements, etc.

The time taken for the ELU to process licence applications varies. On average it takes about a month for a specific licence, and up to three months for some OIELs where several different countries of destination are involved. When an emergency export has arisen at very short notice, the ELU have always responded well when satisfied as to genuine urgency and the unforeseen nature of the circumstances. Responses to inquiries into any unusual delay by the DTI in processing licensing applications generally indicate that the delay arises outside the ELU in the 'advising department' – usually the MOD. Offers to assist the 'advising officer' concerned with the provision of any further information that may resolve any questions and doubts are declined by the DTI, which also will not reveal where exactly within the MOD the application has been referred. The DTI reluctance to inquire further or disturb the MOD in their deliberations can be frustrating, particularly when the DESO have already blessed the proposed export, the contractual date for shipment is close, and the destination country is not one to which 'special export licensing procedures' apply – that is, a country where risks of diversion/proliferation are not normally present.

Clearly the practice of export controls in the UK is complicated, more complicated perhaps than most commentators realize. Problems facing exporters in operating these controls include:

(1) Unfamiliarity with export licensing law and related procedures on the part of those within very large organizations who are not immediately concerned with the actual movement of licensable goods overseas. Highly motivated staff involved in design, research and development, production, publicity, sales, overseas exhibitions, customer relations, product support, and so on may not have export licensing control foremost in their minds when discussing with/providing information to/facilitating urgent deliveries to overseas customers; in practice staff who arrange the actual shipment of export goods assume the responsibility of ensuring that any necessary licence/waiver is in place prior to shipment. Certainly this is the position within BAe, where an initiative has been taken by the Military Aircraft Shipping Department (recently renamed Export/Import Control) to raise awareness in all other BAe departments of the DTI requirements and of the procedures to be followed when controlled goods/technology are dispatched to the EC or exported to non-EC destinations – whether shipped as commercial freight, sent by courier, or hand-carried in personal baggage by BAe technicians;

(2) Uncertain liability to export control, that is, whether or not an item falls within the technical specification of one or more headings of the Control Order. This has been largely overcome within BAe's defence companies by regarding doubtful Group 3 items (dual-use) as being caught by Group 1. Many aircraft structural components of a required specification are of course equally capable of general engineering use, and are purchased by BAe 'off the shelf' from suppliers able to meet the set specification. Such items are not 'specially designed or modified for military use' and appear to be outside the scope of export licensing control. Nevertheless such goods are treated as licensable when dispatched to overseas customers. This reduces the overall control problem for BAe, for the DTI

and also for customs at the point of dispatch/export. But there have been a number of problems with the former Yugoslavia; for example in 1993 a consignment of 190 Land Rovers for Croatia was hurriedly blocked by the DTI, despite two earlier written assurances that the goods were not licensable;

(3) On occasion, the control of old technology: for example, just prior to German unification and before publication of the OGEL (Military Technology), BAe received a request for copy plans of the wings of a Sopwith aircraft (*circa* 1913) from an East German museum engaged in the reconstruction of such an aircraft. The EG(C)O does not concern itself with the age of military technology and after consulting the ELU an export licence had to be applied for, and was eventually granted. Restrictions on the export of spare parts for more recent military aircraft (now deemed obsolete/obsolescent) can highlight some inflexibility in the control system, which can be burdensome both for industry and for the DTI;

(4) The export of damaged or non-operational components of military aircraft to the overseas manufacturer or vendor for repair and return. If components of BAe aircraft are imported for repair in the UK, they can be re-exported after repair to the country from which they were imported, without the requirement of an export licence. However, if such goods cannot be repaired in the UK and have to be sent in their damaged/non-operational condition to the original overseas manufacturer or vendor for repair and return, a licence is required. This disparity in treatment ignores the difference between the positive security risk attached to the operational item returning to its user and the minimal risk of a non-operational item returned to its technology source. Furthermore, no regard is paid to the fact that in the latter case both the aircraft manufacturer and user have agreed that the item is defective – no longer capable of the use for which it was designed – and arguably, therefore, outside the scope of Export Licensing Control. Also government policies and attitudes towards a foreign customer can change very rapidly. (If BAe gets something back for repair from a customer who suddenly becomes an enemy, it is in trouble.) This general problem has been referred by BAe to the ELU several times and still remains unresolved;

(5) Minimal changes in UK licensing procedure arising from the establishment of the EC Single Market at the beginning of 1993. This has

> not so far had any major impact on the DTI system of export licensing. Additional documentation must now accompany intra-EC movements of arms, but otherwise transfers of licensable goods to member states still have to be licensed as before. The debate on possible adoption of an EC regulation to coordinate national export controls on dual-use goods and technologies continues, and if implemented could lead to freer movement of certain goods within the EC, and to common licensing controls at the national frontiers of member states. But the regulation is not yet in place.

By contrast, significant changes in customs control of intra-EC trade have been made. From 1 January 1993, customs no longer require formal entry and do not perform routine physical checks on intra-EC trade at UK ports and airports. Where the dispatch of licensable goods to the EC is covered by an 'Open General' or 'Open Individual' licence, exporters must simply comply with the conditions specified in the licences and keep records of such exports made for production to the DTI or customs, if required. Where the dispatch of licensable goods to the EC is covered by a specific licence, exporters must normally present the licence to their local Excise and Inland Customs Office prior to dispatch of the goods, unless alternative arrangements have been agreed with that office. Where the dispatch to the EC is covered by an OIEL or OGEL, the shipping documentation must contain a reference to the official number of the OIEL or the name of the OGEL whenever an EC movement is effected under such licences. Where the movement involves military goods or firearms/ammunition within the EC Directive on the 'Control of Acquisition and Possession of Weapons', a copy of the specific or 'Open Individual' licence must accompany the goods.

UK export controls of licensable goods including arms to destinations outside the EC are not affected by the changes arising from the creation of the Single Market.

One other significant new development is that, following a series of consultations with companies and trade associations most concerned with export licensing, the ECO initiated work on guidelines for compliance with export control requirements. The intention was to draw up, in consultation with practitioners of export licensing procedures within companies, a code of best practice which could eventually be applied across all sections and types of organization. This has now been done and copies of the code are available.

Bearing in mind the recent reduction in the overall number of controlled items, the increasing risks of proliferation, the changes in proscribed destinations and the development of the European Single Market, together with the serious threat to political stability growing daily in Eastern Europe, the DTI sees it as essential in the changing licensing scene that there should be a high level of commitment within exporting companies at senior management level to ensure compliance with export control. The main features of the guidelines include a statement of commitment from the senior executive of a company to implement the necessary procedures to ensure export control. The draft statement to be made by the Chairman of BAe is reproduced in Annex A, at the end of this chapter. Clear identification of responsible personnel charged with the effectiveness of the company procedures is necessary for DTI audit purposes (and no doubt legal action, if appropriate). Facilities for information and training are considered to be vital, in conjunction with detailed in-house procedures at each stage from contract to delivery, and action to be taken if suspicious inquiries or orders are received.

Conclusion

The DTI move to involve industry in the consultation process prior to finalizing the core of the guidelines is a welcome gesture, intended perhaps to secure the prospect of acceptance by industry prior to implementation and possibly increase the prospects of compliance thereafter. BAe endorses this approach, and from involvement in meetings with the DTI on this subject the company can see no objection to its purpose, or any significant implications for BAe resources.

The proposition that firms should be held responsible – at least in part – for the end-use of their dual-use exports is a reasonable one and interestingly has been referred to in the recent report by the Japanese Security Export Control Committee Industrial Structure Council to MITI, which is reproduced in Appendix 2.

It is unrealistic to expect a licensing authority and its advisers to be aware of all potential uses of all products, by all consignees, within a growing list of sensitive-destination countries. In the course of market research, pre-contract discussions and negotiations, and the commercial vetting of potential customers, opportunities are presented to the exporter to obtain some confirmation of ostensible end-use. Particularly if any doubts or suspicions arise, the exporter should properly inform the licensing authority vis-à-vis the action currently required for goods specified in Part II of the Schedule to the EG(C)O.

Nevertheless, exporters should not be held solely and totally accountable for any misuse subsequently discovered, provided they can demonstrate that all reasonable inquiries as to bona fide use were satisfactory prior to export. Standard 'non-disposal' clauses are contained in all our military aircraft contracts with overseas customers, which can serve as a basis for inquiry into or discussion of end-use in the pre-contract stage.

Finally, one aspect of control which has not been a problem for BAe but where some improvement might be considered concerns DTI supervision of and liaison with customs, which essentially perform an agency function for the DTI in the exercise of physical controls at the UK frontier and in the investigation/prosecution of suspected licensing offences. Not infrequently, customs investigators are prone to take the view that licensing offences have been committed with 'intent to evade the restriction'. Without reference to the DTI, they proceed with a criminal investigation – sometimes with disproportionate zeal – which can have disastrous consequences for exporters (such as Walter Somers, Sheffield Forgemasters, Matrix Churchill) who believe the exports in question have been sanctioned by the DTI. This has led to the inquiry into the export of arms to Iraq headed by Lord Justice Scott. On some occasions, customs officers have seized export consignments and on appeal to the courts, exporters have been able to establish – after protracted delay by customs in initiating the necessary 'condemnation proceedings' – that in fact the power of seizure was not available to customs in the circumstances of the alleged offence. At present, years can elapse before exporters can obtain even partial recovery from customs of the total costs incurred by improper seizure. A time limit of one month from the date of presentation of the relevant entry is placed upon appellants (importers/exporters) within which to appeal against seizure (Schedule 3 CEMA 1979). But there is no time limit placed upon customs to initiate legal proceedings to test the validity of the appellant's claim. Customs habitually ignore any representations or explanations offered by 'offenders' and have to be continually pressed to bring the matter before the courts. A suitable amendment to Schedule 3 is long overdue, requiring customs to apply for a civil hearing in the appropriate Magistrates'/High Court within a reasonable period of the date of formal appeal in seizure cases. Perhaps in the interests of overall export control compliance and best practice, it would be useful to include customs headquarters in the present joint industry/DTI discussions.

However, this is essentially a technical problem of UK bureaucracy. Behind the practice of export controls in any country lie fundamental political and technical problems:

(1) how to define dual-use technology;
(2) how to harmonize differing national standards in order to achieve fair play;
(3) how to make sure the controls are effective; and above all,
(4) how to define your enemies.

Notes

1 W. Cunningham, *Growth of English Industry and Commerce* (Cambridge: Cambridge University Press, 1905), i, pp. 433–4.
2 T. H. Lloyd, *England and the German Hanse 1157–1611* (Cambridge: Cambridge University Press, 1991), p. 49.
3 The Alabama case concerned the arming of a Confederate ship by British private interests during the American Civil War in the 1860s. It led to the enactment of the Foreign Enlistment Act 1870, which put controls on the activities of British subjects going abroad as mercenaries.
3 Anthony Sampson, *The Arms Bazaar* (London: Hodder & Stoughton, 1977), p. 40.
4 J. D. Scott, *Vickers* (London: Weidenfeld & Nicolson, 1962), pp. 254–6.
5 SIPRI annual reports, *passim*.
6 Personal recollection by Noel Johnson to author.
7 Cf. comment by Sir R. Smith in interview on the BBC TV programme *Midweek*, 2 July 1975, quoted by Sampson (op. cit., p. 13): 'We recognise arms as not a means to starting a war, but as a means to stopping a war' (p. 13). The question of lifting the arms embargo to Bosnia involves the same issue.
8 Ian Anthony, in I. Anthony, ed., *Arms Export Regulations* (Oxford: Oxford University Press, 1991), p. 9.
9 House of Commons Trade and Industry Committee, Session 1991–2, Exports to Iraq, London, HMSO, 1992, p. vii.

Annex A: Draft: BAe Chairman's statement of 'virtue'

As a major exporter of advanced technology the Company needs to be constantly aware of export control regulations and to set high standards of compliance. Failure to comply can result in severe penalties for the Company and individual employees.

The purpose of export control regulations is to restrict the access of certain countries to strategic equipment and technology capable of being used for military or terrorist activities. The regulations under UK law form part of a wider international control effort. They bear particularly on the Company's defence and space businesses, and others, such as civil aircraft and motor vehicles.

The Department of Trade and Industry, which is responsible in the UK for matters affecting export controls, recently issued a Code of Practice designed to assist industry in meeting the levels of compliance required under these regulations. The Code was drawn up in consultation with industry, including representatives from BAe, and is now being made available to all of the Company's business units. Although the Code is voluntary, and therefore does not have the force of law, it does represent 'best practice' which the Board has resolved to follow.

Responsibility for implementing the Code of Practice within each business unit rests with individual Managing Directors. They will initiate a review of their procedures in the light of the Code and ensure that the correct training and other measures are taken so that the Company maintains the standards of compliance which the Board has established.

APPENDIX 1

MAJOR CHANGES IN GLOBAL SECURITY EXPORT CONTROLS AND THE ROLE OF JAPAN

OSAMU WATANABE

We are pleased to include this address by Osamu Watanabe, Director General of the Ministry of International Trade and Industry, as a statement reflecting official thinking and policy in Tokyo. It was given at the dinner of the third conference organized by the Royal Institute of International Affairs and the Institute for International Policy Studies, in June 1993.

Introduction

The global security export control issue is now at a turning point, and I consider this conference in Tokyo to be very timely. Today, I would like to present my views on the present state of global security export controls and on the way Japan should respond.

International trends in security export controls

First of all, I would like to review the major changes in global security export controls in recent years.

Changes in the security environment

The end of the Cold War, which dominated the international political scene for so long, has necessitated a fundamental change in security export controls.

There are now growing concerns about the emergence or aggravation of regional disputes caused by territorial claims, resource availability, and racial and religious conflicts that had been suppressed during the Cold War by the presence of the two superpowers. Such concerns have become reality in some regions, as seen in the Iraqi attack on Kuwait and regional strife in the former Yugoslavia. It is also feared that these regional disputes may become linked to the proliferation of weapons of mass destruction, that is nuclear, chemical, biological weapons and their delivery systems missiles, leading to further regional instability. This danger became especially apparent in the Middle East after the Cold War and during the Gulf war.

In the face of these trends, the prevention of the proliferation of weapons of mass destruction and their delivery systems, and the prevention of the excessive build-up of conventional arms, have become major global security concerns. Consequently, the importance of the export controls is now internationally recognized.

A strengthened control of the proliferation of weapons of mass destruction and recent approaches of the Japanese government

The export control regimes for preventing the proliferation of weapons of mass destruction and missiles have recently been strengthened and enhanced through the activities of the Nuclear Suppliers Group, the Australia Group and the MTCR. In this context, Japan has been playing a positive role in the global security regimes.

Japan took over the secretariat of the Nuclear Suppliers Group with the agreement of the group concerning export restrictions on nuclear-related dual-use items in April of last year. Japan is taking the initiative in this group. Non-proliferation export controls require the confirmation of end-use and end-users. However, there are differences in such licensing procedures of export controls among the member countries of some regimes. In order to deal with this problem, Japan is making concrete proposals in each regime to harmonize the licensing procedures of the member countries and thus to enhance the effectiveness of export controls.

Japan's global activities in export control regimes have been fully recognized by other participating countries.

Future problems in the global security export controls and Japanese approaches

Next, I would like to discuss five changes that are important for the future of global security export controls.

1. Coordination of the non-proliferation export control regimes

First, the best form of non-proliferation export control regimes must be studied from a long-term point of view and on a global basis.

Japan should then make proposals in the international arena on ways of coordinating the existing export control regimes of nuclear, biological, chemical weapons and missiles, which to date have only been discussed within separate frameworks. Japan may also suggest ways of unifying these regimes to form a system for making unified decisions. In discussing these issues, we must consider how the whole of non-proliferation export control regimes should be reorganized from a comprehensive and long-term point of view. These discussions must also take into account such major trends as the NPT review and changes in COCOM that have occurred as a result of the end of the Cold War.

With these issues in mind, in addition to harmonizing the licensing procedures of member countries in each regime, Japan is considering coordinating the differences among the regimes. More concretely, given the common nature of the regimes as to the regulations on dual-use items of weapons of mass destruction, we believe that the lists of controlled goods, licensing procedures and approaches to non-member countries can be coordinated among the regimes. We also believe that such a coordination effort will enhance the effectiveness of the non-proliferation export controls in total.

2. Helping to refine other nations' export controls

Second, advanced countries should help other countries to establish effective export control systems.

Since many states in the former Soviet Union, Eastern Europe and Asia are now capable of supplying controlled goods, it is necessary to have their cooperation so that international security export controls can be more effective. To this end, we must try to have these countries understand the basic principles of security and to induce them to cooperate under the export control regimes, through such measures as sending specialist missions, holding seminars and offering training courses.

Such missions and seminars have already been started for the states of the former Soviet Union and East European countries to some extent. Examples include last year's Warsaw Seminar and Budapest Seminar, which were held by the MTCR and the Australia Group respectively. Japan also invited experts from Russia to lecture on export controls this February.

However, the very rapid economic growth of Asian countries, and the consequent possibility of their becoming major suppliers of industrial products, suggest that we must attach greater importance to this region.

From this point of view, we plan to hold an Asian Export Control Seminar from 25 to 27 October 1993 and intend to invite the policy-making officials of the six ASEAN countries, the Republic of Korea and Hong Kong. With this seminar as a starting point, Japan should work to build a cooperative relationship with these Asian countries regarding export controls. Several developed countries such as the United States and Australia have already expressed positive support for our idea. Actually, this seminar will be held in cooperation with some advanced countries. It is essential to make security export controls truly global through these efforts.

3. A changing COCOM

A third area that is important for the future of the global security export controls is the transformation of COCOM.

With the end of the Cold War, COCOM is about to end its role of restricting the outflow of strategic commodities to the communist bloc for the sake of safeguarding the Western world. However, its new function has yet to be conceived.

After the Secretary of State Jim Baker of the US had made a proposal for COCOM's transformation in June 1992, COCOM started to change. The first COCOM Cooperation Forum was held in November 1992 as an opportunity for dialogue between COCOM member countries and proscribed countries.

The birth of such a forum where COCOM countries and proscribed countries can cooperate against the threat of countries of concern indicates a major turning point of COCOM from 'confrontation' to 'collaboration'.

However, the Cooperation Forum itself is just the start of the change in COCOM. We are now faced with the questions of what new direction COCOM should take and how we can properly prepare for the threats to global security. As my personal views on the future changes of COCOM, I would like to make the following three points. First, COCOM controls should be relaxed further in response to democratization and liberalization of proscribed countries. Second, as a condition for such deregulation, effective export controls should be established in these countries. Third, through these changes, we should respond to new regional threats in cooperation with those proscribed countries.

4. Export controls on conventional arms

The fourth issue, which is critical for global security, is that of export controls on conventional weapons.

With the intensified regional conflicts following the end of the Cold War, the export controls on conventional arms have become a key issue. The supply of conventional arms from surrounding countries to the former Yugoslavia and Cambodia is aggravating the conflicts. Meanwhile, the pressure to export weapons is increasing in the former communist bloc, especially in Russia and in China.

Such weapons should be controlled under international coordination because the transfer of conventional arms to the countries and regions where conflicts are developing and expanding can threaten global security. Based on this concept, the United Kingdom, Japan and other countries proposed a registration system of conventional arms to the UN in 1991. The aim of this system is to improve the transparency of the transfer of conventional arms and make the records more accessible. The proposed system was implemented in January 1992, and we look forward to seeing it operate effectively with support from an increasing number of participating countries.

Also, in order to prevent excessive build-up of conventional arms, the G7 have been holding international discussions on export controls of dual-use items related to conventional arms. However, such controls should be supplementary to the control of transfers of conventional arms themselves to countries of concern. In other words, it is essential to restrict the transfer of the conventional arms. Japan has strictly restricted the export of arms, based on the 'Three Principles' of arms exports. Japan will require other countries to take a cautious approach regarding the export of conventional arms, so as not to cause regional instability through the increase of such exports in the world.

5. Reinforcing Japan's export controls

Finally, I would like to present my view on Japan's export controls.

The discovery after the Gulf war of Iraq's efforts to develop nuclear weapons has increased awareness that goods other than internationally controlled items may be diverted to the development and manufacture of weapons of mass destruction. To prevent such a scenario, the United Kingdom, the United States and Germany have already introduced export controls using the 'know' standard.

In response to such new regulations in other advanced countries, the Industrial Structure Council of MITI established the Security Export Control Committee to study the future of export controls. Honourable Mr Imai, who is attending this meeting, played an indispensable role as its chairman. Through the intensive discussions in this committee, we obtained the conclusion that, in cooperation with other advanced countries, Japan should introduce the export controls on goods other than internationally controlled items.

In its final report, the committee says that, when applying export controls to such a wide variety of goods, restrictions should be placed only on the products that have possibilities of being used in the development and manufacture of weapons of mass destruction. This is to avoid placing excess burdens on exporters and to avoid restricting the economic growth of developing countries.

The committee also says that with a view to implementing the necessary and minimal controls, it is appropriate to limit the list of countries subject to controls and to refrain from controlling goods that have little possibility of materially contributing to the development and manufacture of weapons of mass destruction. Based on this report, we intend to draw up the outline of the new export controls around autumn 1993, and to start to implement it from the beginning of 1994.

It is essential to ensure international effectiveness in introducing such new export controls. Therefore, we should seek international harmonization in this field as well.

Conclusion

The international situation regarding security export controls is now at a major turning point. The importance of non-proliferation export controls has been reaffirmed by the Gulf war. Now a comprehensive new design for non-proliferation export controls is required. In order to implement the design, it is essential to advance international collaboration among such countries as the United Kingdom and Japan, and to take into account the opinions not only of governments, but also of academics. The wide-ranging discussions at this conference, the third in the ongoing RIIA/IIGP project, conform to the objectives of non-proliferation controls, and are useful in considering the way the global security export controls ought to be in the future.

APPENDIX 2
THE FUTURE OF SECURITY EXPORT CONTROLS

Unofficial translation, 25 March 1993. Security Export Control Committee, Industrial Structure Council, Ministry of International Trade and Industry

PREFACE

The Western Free World has enjoyed post-war economic growth in the free trade system under the GATT (General Agreement on Tariffs and Trade). This in turn is based on the existence of a free and stable international community. Thus, the GATT, while devoted to the development of the free trade system, also recognizes the need for an exception for controls for security purposes. This recognition has also been the basis of the COCOM controls implemented by the Western Free World.

Japan has also implemented export controls for security purposes during the four decades since joining COCOM in 1952. The rapid economic development that Japan achieved over this period has caused a corresponding increase in its responsibilities in security export controls. Japan has responded to this challenge, and contributed to the development of a free society as a member of the Western Free World, by gaining experience in security export controls and building an elaborate export control system.

However, the end of the Cold War, which defined the postwar international order, demands a complete international overhaul of export controls. The disintegration of the Soviet Union calls for a rethinking of the significance of COCOM controls, which was intended to restrict the outflow of strategic commodities from the Western Free World to the Communist Bloc. The Gulf War has re-emphasized in the public mind the threats that the proliferation of weapons of mass destruction and the excessive build-up of conventional arms pose to global security. It is in the face of these changes in the security environment that efforts are being made to transform COCOM and reinforce non-proliferation export controls. Security export controls are at a major turning point.

As a major producer and exporter of high technology products, Japan's responsibilities to global security through its export controls are great indeed. Reinforcing export controls and other efforts in its economic activities to achieve world peace are an area in which it is most appropriate for Japan to play a role commensurate with its status as a major economic power.

There is a need to develop an export control policy in Japan that meets these changes in the international environment and befits its role in international politics, based on a fundamental rethinking of its current policy.

The Industrial Structure Council established the Security Export Control Committee in September 1992, and conducted the first intensive debate in Japan on this issue. This report makes proposals on the future of security export controls, based on this debate.

I. INTERNATIONAL TRENDS SURROUNDING SECURITY EXPORT CONTROLS

1. Changes in the security environment

International developments on security issues since the end of the Second World War evolved within a 'stable' global order, i.e. the Cold War, structured around the United States and the Soviet Union, the two superpowers with vast differences in political and economic structures.

With the progress in reforms in the Soviet Union and Eastern Europe, starting with *perestroika*, the East–West relationship has changed from confrontation to cooperation. Western countries are providing assistance to the Eastern countries in their move towards democracy and the market economy. In this process, the dissolution of the Soviet Union, one of the two main protagonists of the Cold War, in late 1991, brought a definite end to the Cold War structure, causing major changes in the international political scene.

Specifically, there are concerns about the emergence or aggravation of regional disputes concerning territorial claims, resource availability, ethnic conflicts, and religious antagonism, that had been suppressed during the

Cold War by the presence of the two superpowers, as new threats to global peace and security. Such concerns are becoming reality in some regions, including the Iraqi attack on Kuwait, and disputes in former Yugoslavia and the former Soviet Union.

The Gulf War, and other developments in the Middle East in the post-Cold War era have brought to the surface moves to possess weapons of mass destruction, i.e. nuclear, chemical and biological weapons, and their delivery systems, i.e. missiles, and rapidly to build up conventional arms. It is feared that these trends will be linked to regional disputes, leading to further regional instability.

In the face of these trends, international efforts are under way to build cooperative relationships to prevent the proliferation of weapons of mass destruction and missiles, and the excessive build-up of conventional arms.

2. Transformation of COCOM

The traditional role of COCOM was to ensure the security of the Western Free World by preventing the outflow of strategic commodities to the former Communist bloc. This role is disappearing with the end of the Cold War, and efforts are under way to deal with a new issue: responding to new regional threats.

In May 1992, the United States proposed that COCOM member countries and proscribed countries cooperate in dealing with new strategic threats. This proposal was taken up in November 1992, at the first meeting of the COCOM Cooperation Forum, which afforded an opportunity for dialogue between COCOM member countries and proscribed countries. Such attempts to adapt COCOM to a new strategic environment through cooperation between member countries and proscribed countries indicate a major turning point for COCOM.

3. Reinforcement of multilateral export control regimes

There has been a renewed awareness after the Gulf War that the proliferation of weapons of mass destruction and missiles and the excessive build-up of conventional arms pose a threat to global security, and that appropriate export controls for dual-use items which may contribute to this are important in meeting this threat. This understanding was confirmed, for example, in the Political Declaration at the Munich Summit in 1992. The last few years have seen the rapid development and reinforcement of export control regimes through the expansion of their membership and controlled items.

(1) Nuclear weapons

The NPT (Treaty on the Non-Proliferation of Nuclear Weapons), with 156 parties, and the NSG (Nuclear Suppliers Group: 26 nuclear supplier countries) Guidelines have provided the framework for export controls on nuclear related goods and technologies. The accession of China and France to the NPT in 1992 brought all Nuclear Weapon States as defined in the NPT under its aegis.

In addition, it was agreed at the NSG meeting in April 1992 to control the export of dual-use goods and technologies which can be used for the development and manufacture of nuclear weapons. Japan is taking the initiative in developing international guidelines for this purpose, and has agreed to act as the point of contact.

(2) Chemical weapons

As for chemical weapons themselves, the Gulf War spurred the move towards an early agreement on the Chemical Weapons Convention. In 1992, the UN General Assembly adopted a resolution recommending the signature and ratification of the Convention. More than 130 countries have already become signatories. A major feature of this Convention is its provision for wide-ranging and strict verification measures, including challenge inspections.

Export controls on chemical weapon precursors have been implemented since 1985 through the Australia Group, consisting of the major suppliers. In 1992, chemical weapons-related equipment also came under control. The membership of the Australia Group has also been expanded, from the original 15 to the current 24 countries.

(3) Biological weapons

The Biological Weapons Convention has been in force since 1975. In addition, in December 1992, the Australia Group agreed on the lists of human pathogens, animal pathogens, and biological dual-use equipment.

(4) Missiles

Export controls on delivery systems for nuclear weapons and related equipment and technologies have been imposed since 1987 under the MTCR (Missile Technology Control Regime), consisting of the major suppliers. Membership has expanded to 23 countries. After the Gulf War, in 1992, the restrictions were expanded to cover any delivery system for weapons of mass destruction, including chemical and biological weapons.

(5) Conventional arms

In 1991, the five major suppliers of conventional arms, the United States, Russia, China, the United Kingdom, and France, conducted discussions, and, in October of the same year, agreed on the Guidelines for Conventional Arms Transfers. The UN General Assembly in its 1991 Plenary Session adopted a resolution for the establishment of the UN Register of Conventional Arms. In addition, in November 1992, the G7 agreed to apply export controls on dual-use items which could be used for the production of conventional arms in Iran, Iraq, Libya and North Korea, to which the G7 had already forbidden the export of conventional arms themselves. Discussions are continuing to enhance the effectiveness of such controls.

4. Reinforcement of export controls in developed countries and assistance to other countries

(1) Reinforcement of export controls in developed countries

Based on the international recognition that the prevention of the proliferation of weapons of mass destruction and missiles is essential to global peace, developed countries, in addition to implementing export controls based on international agreements, are undertaking measures on their own to reinforce security export controls, to fully control exports from a security point of view.

The United States, the United Kingdom and Germany have added items which may contribute to the development or production of weapons of mass destruction and missiles to their lists of controlled items.

These countries have also introduced export controls using the 'know' standard, and other countries such as Australia are also considering such introduction. Controls using the 'know' standard subject exports of goods other than those on their lists of controlled items to approval by administrative authorities when the exporter 'knows' that they will be used to develop or produce weapons of mass destruction or missiles.

They are also taking measures to develop security export control systems within firms. Germany, the United States and the United Kingdom have announced guidelines for export controls within firms. Germany requires firms to designate a person in charge of security export controls.

There are other efforts to strengthen the administration of export controls, such as reinforcing the relevant institutions (Germany has established a Federal Export Office), and increasing security export control personnel. Germany has taken further drastic measures to strengthen security export controls. It has twice increased prison terms, fines and other sanctions on unapproved exports, and beefed up monitoring by giving the federal agency for export control and customs investigations authority to open letters and other mail and to conduct wiretapping.

(2) Assistance to other countries to establish their security export control systems

The expansion of export controls to a wide range of items, together with industrial development, has brought states, such as those in the former Soviet Union, Eastern Europe and Asia, into the scope of supplier countries of controlled goods. This has led to the recognition of the need to establish security export control systems in these countries in order to implement internationally effective security export controls on a cooperative basis.

Japan, the United States, the United Kingdom, Germany, Australia and other developed countries have given assistance to countries in the former Soviet Union and Eastern Europe, and countries in Asia such as the Republic of Korea, Vietnam, Mongolia, in the establishment of their security export control systems. Hungary and other countries receiving this assistance have shown growing awareness of the importance of security export controls, making initial efforts to establish security export control systems and participating in multilateral security export control regimes.

II. JAPAN'S SECURITY EXPORT CONTROLS: CURRENT SITUATION AND ISSUES

1. The current situation of Japan's security export controls

Security export controls are an exception to free trade. Since joining COCOM in 1952, Japan has exercised export controls on COCOM controlled goods and technology. It has also imposed strict controls on the export of arms.

The Toshiba Machinery Co. case in 1987 created a renewed awareness of the importance of export controls from a security viewpoint. This resulted in the amendment of the Foreign Exchange and Foreign Trade Control Law to fortify criminal and administrative sanctions on violations, the reinforcement of export controls administration, and other measures to strengthen the export control system. CISTEC (Center for Information on Strategic Technology) was established as the main institution for security export controls in the industrial sector, and has worked to raise awareness of security export controls and disseminate relevant information. In response to a request from the Minister of International Trade and Industry for the strict enforcement of export control laws and regulations, the industrial sector has worked to develop in-house control systems through such measures as the establishment of compliance programs. Such efforts and experience gathered over the past four decades have led to the establishment of a system, including the development of security export control experts in the public and private sectors, which provides export controls that meet international standards.

There is recent wide recognition in Japan of the importance of non-proliferation export controls, that is leading to active ongoing efforts in this respect.

2. Future issues

(1) The need for detailed and sophisticated controls

COCOM controls could achieve their objective if all the exports of proscribed items to proscribed countries were banned in principle, regardless of the end use of goods to be exported. If this retarded economic development in the proscribed countries, this was accepted as an inevitable side effect.

On the other hand, non-proliferation export controls are aimed at preventing the proliferation of weapons of mass destruction and missiles, and the excessive build-up of conventional arms. If such export controls obstruct economic development of countries subject to control, this will invite their economic isolation, which could actually be a negative factor for regional security. Therefore, non-proliferation export controls should be implemented with restraint so that they will achieve their objective, and should not retard the economic development of countries subject to control.

Thus, in implementing non-proliferation export controls, unlike COCOM controls, it is necessary to confirm the end use of goods to be exported, which in turn calls for detailed and sophisticated controls.

However, when the government implements such export controls, activities beyond the border where Japanese authority cannot reach, must be taken into consideration in making decisions. This is making it more difficult to enforce controls. Exporters also face various difficulties in checking on the business activities of the recipients of the exports.

(2) The renewed need for international cooperation

In COCOM controls, there is a clear and detailed understanding on the implementation of controls, covering proscribed countries and items and procedures, and a well-established secretariat.

However, non-proliferation export controls, which are seeing recent and rapid expansion, are implemented through cooperation among a large number of countries to deal with the risk of proliferation of weapons of mass destruction and missiles and the excessive build-up of conventional arms, and do not reflect ideological confrontation like COCOM controls. Therefore, the positions of participating countries may diverge because of differences in their perceptions of countries of concern, and their diplomacy. As a result, there is insufficient international agreement on the details for implementation, such as the means and criteria for determining that the goods in question will not be used to develop or produce weapons of mass destruction and missiles. Unlike COCOM controls, this has created broad areas where each country must exercise its own judgment. Thus, without international cooperation on coordination and harmonization, such as information sharing among member countries and the development of common international guidelines on the implementation of export controls, there is the danger that the controls will not be fully effective.

(3) The need to expand the scope of export controls

Traditional export controls by developed countries have generally established lists of controlled items through international agreements. However, since the IAEA inspections after the Gulf War disclosed Iraqi efforts to develop nuclear weapons, there is a growing international consensus over the need to place goods other than internationally controlled items under export controls. In the United States, the United Kingdom and Germany, this has already led to the implementation of controls on the export of such goods. Other countries such as Australia are also preparing to introduce such measures. Since Japan is a major supplier of manufactured products, there is a concern that its products may be used to develop and produce weapons of mass destruction and missiles, which is creating a growing awareness that Japan must act expeditiously to take new measures in addition to existing controls.

III. FUTURE SECURITY EXPORT CONTROL POLICY

1. The role of security export control policy and its significance to Japan

(1) The role of security export controls in overall international security policy

Maintaining global peace and security requires comprehensive efforts in such areas as strengthening the role of the UN, developing regional security frameworks, diplomatic efforts such as the promotion of dialogue between parties to regional disputes, eliminating destabilizing factors through economic cooperation and other efforts, and arms control and disarmament. Export controls for security purposes are an important part of arms control. However, they are to be used in unison with other policies to achieve peace, and should not be considered a policy that can by itself completely prevent the proliferation of weapons of mass destruction and missiles, or the excessive build-up of conventional arms.

Besides, export controls for these purposes, unlike economic sanctions, are not aimed at constraining ordinary economic activities in the countries subject to export controls, but are intended to prevent the possession of weapons of mass destruction and missiles or the excessive build-up of conventional arms. Indeed, it is often important to prevent their isolation and eliminate regional destabilizing factors through economic development in these countries. In some eases, it may be appropriate to extend technical and economic cooperation to such countries. It is necessary in implementing security export controls to make due considerations so that obstacles to healthy economic development will be minimized.

(2) The significance of security export controls to Japan

As a major producer and exporter of manufactured goods, Japan is facing an increasing need to build a more effective security export control system that will prevent the use of Japanese products in the development or production of weapons of mass destruction and missiles. This responsibility is evident, in view of Japan's reliance on a free and stable international community.

Japan has imposed strict controls on the export of strategic goods. It is necessary for Japan to continue its efforts to implement effective security export controls. Based on such achievements and efforts, Japan must coordinate its security export control policy with other security policies, and call on the cooperation of the international community in security export controls and seek their understanding and consensus. This is the main role that Japan must play.

2. Japan's role in the international community

(1) Coordination of multilateral security export control regimes

In order for controls to be effective, supplier countries must implement them on the basis of a common understanding and common principles. This calls for efforts to share security-related information through international exchange, and to develop clear guidelines for the international harmonization of their implementation.

With the agreement on biological weapon-related agents and equipment in December 1992, the control mechanism for weapons of mass destruction and missiles now covers all areas. However, since controls related to nuclear weapons, biological and chemical weapons and missiles have been discussed under different regimes, problems have surfaced, such as the differences in the elaboration of the guidelines, as well as membership.

In order to promote cooperation essential to the effectiveness of international export controls, Japan must play an active role in promoting harmonization and coordination among the various regimes, including sharing information on and evaluation of threats posed by countries of concern and other security issues.

The consolidation of existing security export control regimes should also be explored, with the aim of deepening common understanding on security issues including evaluation of countries of concern and the promotion of more effective harmonization and coordination in various export control areas.

(2) Promotion of international controls on transfer of conventional arms

There are various difficulties in controlling the export of conventional arms. Unlike weapons of mass destruction, whose production is, in principle, internationally recognized as being banned under treaties, some say that it is inappropriate to restrict the export of conventional arms across the board, since controls will directly affect the right to self-defence, as well as security, of the countries subjected to constraints on their access to conventional arms. Moreover, supplier countries of conventional arms consider export controls on such arms to be a matter of their own discretion, as an important part of their national security policy. Because of this, in controlling the transfer of conventional arms in the post-Cold War world, although efforts are under way to increase transparency through the newly established UN Register of Conventional Arms, there still is no international agreement on a means of restricting the transfer of them.

However, the transfer of conventional arms to countries and regions, where there are concerns about new disputes and the aggravation of existing ones, could pose threats to global security. There are various international moves to prevent this. It is desirable to build on this, and control the transfer of conventional arms in some form of international cooperation. Moreover, although export controls on dual-use goods related to conventional arms have been discussed within the G7, this has meaning only if the transfer of conventional arms themselves is internationally controlled. Japan should call on other countries for international cooperation in strengthening the control of conventional arms exports to countries of concern.

(3) Assistance to other countries to establish their export control systems

In building an effective international framework for security export controls, it is important to ensure the participation of countries in the former Soviet Union, Eastern Europe and Asia. To this end, it is important to seek their basic understanding of security issues, and encourage their development of security export control systems and their participation in multilateral export control regimes, by dispatching expert missions, holding seminars, accepting trainees and other means.

In the former Soviet Union in particular, the outflow of weapons continues because of the domestic situation, including falling domestic demand for weapons, delays in the conversion from military to civilian production, and the desire to earn foreign exchange. There are concerns that this will be a destabilizing factor for global security. As one response to this state of events, there is an urgent need to establish security export controls in the former Soviet Union. Japan should also actively assist the development of such controls in those countries.

Asian countries, led by NIES (newly industrialized economies) and ASEAN (Association of Southeast Asian Nations) member countries, are enjoying rapid economic growth, and are expected to become major suppliers of manufactured products. Thus, there is an urgent need for their participation in the international security export control regimes. In this respect, it is important for Japan to seek their full understanding of the necessity and desirability of security export controls and encourage their participation in the international security export control regimes. Japan must actively cooperate in helping them rapidly establish their own control systems which is a prerequisite to this participation. We believe that strengthening political and economic ties in Asia through such efforts will contribute to enhancing security in this region.

3. Reinforcing Japan' s export control system

(1) Future export controls

(A) Introduction of new export controls for the prevention of the proliferation of weapons of mass destruction and missiles. The IAEA inspections after the Gulf War and other events have led to growing international concerns that goods other than internationally controlled items will be used to develop or produce weapons of mass destruction and missiles. Japan, as a developed country, must cooperate with other countries to implement export controls for such items, so that they will not be used to develop and produce weapons of mass destruction and missiles.

Implementing export controls for such items by the traditional export control method, i.e. subjecting all exports of controlled items to approval by the administrative authorities, would require exporters to obtain approval for an enormous amount of goods directed to controlled destinations. Since this would impose an insupportable burden on exporters, it is necessary to limit the subject of controls. To this end, it is necessary to introduce a new method of controls, which subjects to controls only cases where there is a possibility that such goods will contribute to the development or production of weapons of mass destruction and missiles.

(a) limiting controlled destinations and items. In implementing such controls, it is appropriate to keep them to a minimum by limiting controlled destinations to those where concerns are that there may be activities related to weapons of mass destruction and missiles, and by refraining from controlling items which have little possibility of materially contributing to such development and production.

(b) developing clear and objective standards. In implementing such controls, relying on subjective factors, such as whether the exporter 'knows' or 'has reason to know' that goods to be exported will be used to develop or produce weapons of mass destruction or missiles, may lead to uncertainty for exporters and obstruct normal export activities. This is because it will be unclear to what extent an exporter must confirm usage and other factors to ensure that he can be absolved of responsibility if and when the goods are in fact used to develop or produce weapons of mass destruction or missiles. Therefore, in order to define clearly the responsibility of the exporter it is necessary to develop clear and objective standards to determine whether or not the transaction is subject to controls, based on the possibility of use for the development or production of weapons of mass destruction and missiles.

More specifically, it is essential to set objective standards to determine whether or not the business activities of the end user are related to the development or production of weapons of mass destruction or missiles.

In setting the methods and standards for determining the applicability of controls to individual transactions, it is necessary to conduct full deliberations, based on how transactions are actually conducted.

(c) the need for international coordination. It is also necessary to ensure international fairness and effectiveness through international efforts to harmonize implementation and other aspects of the controls.

(d) assistance to small and medium enterprises. It is desirable to assist small and medium enterprises in gathering security-related information, as well as actively to provide advice, consultation and other assistance in developing their compliance systems.

(B) The implementation of focused export controls by streamlining procedures. In order to make export controls more efficient as a whole, it is necessary to focus export controls by streamlining controls, including existing ones, where possible, and concentrating the available resources in the administrative authorities and firms on highly sensitive cases.

Currently in Japan, all destinations are subject to export controls aimed at the nonproliferation of weapons of mass destruction and missiles. However, there are countries such as the United States which apply simplified procedures for exports to certain countries which are members of international export control regimes. Re-export to countries of concern is very unlikely with regard to exports to countries such as the United States, which implement export controls as strictly as Japan. Focusing controls on exports to countries of true concern should be considered.

As for firms with effective security export control systems, consideration should be given to easing their burden by simplifying export approval procedures, under the condition that they strictly enforce compliance programmes.

Moreover, it is important to improve the streamlining of export licensing procedures and related documents in general.

(2) Further improvement of private and public systems

(A) Improving administration. The increasing importance of non-proliferation export controls is expanding the scope of export controls, and calls for increasingly detailed controls. This requires administrative reinforcement of security export controls. Specifically, it is necessary to enhance efficiency through further use of regional bodies and other means, increasing personnel, and training officials with high degrees of expertise matching the increasing complexity of controls.

In addition, it is important to reinforce ongoing efforts to introduce computer systems in export licensing procedures, in order to raise the efficiency of licensing applications and examinations.

(B) Establishing compliance systems in firms. The effective and efficient implementation of security export controls requires not only efforts to develop the necessary administrative authorities, but also sufficient attention and controls by the firms that are engaged in exports. Specifically, it is necessary for each firm to define clearly the lines of internal responsibility,

such as by designating a person in charge of export controls, and to strengthen in-house examination, auditing and other aspects of its export control system.

Each industry should develop guidelines that fit the specific characteristics of that industry, to build efficient and fair compliance systems applicable to its individual businesses.

To assist these activities, CISTEC should expand such activities as collection and analysis of information, response to requests for individual consultations, and education. Reinforcement of such aspects as personnel and finances is hoped for.

(3) Collecting, analysing and providing information

In non-proliferation export controls, information on security matters, such as development plans abroad for weapons of mass destruction and missiles, the military activities of various countries and sensitive export cases, plays a major role in examinations by authorities and decision-making in firms. It is becoming increasingly important to collect such information from a wide range of sources, and to analyse it, so as to implement appropriate controls.

(A) Collecting, analysing and providing information in the administrative authorities. It is necessary to collect and analyse information necessary to security export controls, including overseas information on weapons of mass destruction and missiles, and to develop effective means of utilizing relevant information gathered by ministries and firms. Consideration should be given to having the authorities use such information in their examinations and in responding to requests from exporters. It is also necessary, however, to give consideration to the existence of limitations to public disclosure because of diplomatic and other considerations.

(B) Collecting and analysing information and education in firms. Some firms, because of their business activities abroad, have opportunities to obtain overseas information necessary to security export controls through their overseas offices and other sources. It is hoped that these firms will actively collect, analyse and effectively use such information.

In order to raise awareness, not only in the export control departments, but also in the sales departments of firms, it is necessary to do such things as enhancing education in firms, and conducting seminars and training through export-related industrial associations.

CONCLUSION

This report proposes the basic philosophy of security export controls and future directions thereof, taking into account global changes in the security environment. We hope that appropriate policies based on the proposals will be implemented expeditiously.

When the government implements security export control policy, it is essential to obtain the understanding and cooperation of the general public, as well as exporters. Close coordination of efforts within the government is also important, including information sharing among the ministries and agencies concerned. Moreover, the government should make efforts to raise the effectiveness of policy measures by fully explaining Japanese security export control policy to other countries, and taking a leading role in international efforts for coordination.

The proposals in this report provide an overview of future security export controls. Needless to say, there is the need for further and more detailed consideration of individual policy measures. Especially in introducing new controls to prevent the proliferation of weapons of mass destruction and missiles, it is necessary to conduct deliberations taking into full consideration the actual situation surrounding transactions, and to provide a sufficient lead-time.

Moreover, it is expected that there will be further, major changes in the international security environment, including the countries of concern. It will be necessary to re-examine security export control policy, recognizing that those changes will alter the premises of such policy.

Finally, we wish to state our hope that this report, the first attempt by the Industrial Structure Council, to focus Japanese thinking on security export controls will act as a catalyst in deepening public understanding of security export controls, and enhancing Japanese security export controls.

SECURITY EXPORT CONTROL COMMITTEE

Kensuke EBATA, Defence Commentator (Japanese Correspondent, *Jane's Defence Weekly*)
Riyako GODAI, Commentator (Freelance Writer)
Yoshihiro HARADA, Managing Director, Toray Industries
Yoshihiko HIROOKA, Managing Director, Sumitomo Chemical Company
Hideo HIROTSU, Executive Vice President, Mitsubishi Heavy Industries
Ryukichi IMAI, Former Ambassador for Arms Control and Disarmament (Chairman)
Toshi KITAMURA, Representative Director and Executive Vice-President, Hitachi
Shinichi KITAOKA, Professor, Faculty of Law and Politics, Rikkyo University
Taiichiro MATSUO, President, Japan Machinery Exporters Association
Hiromichi MIYAZAKI, Former Ambassador to the Federal Republic of Germany
Toshihiko MORITA, Executive Vice President, Sumitomo Corporation
Kinji NAKABAYASHI, President, Nagoya Tsusho Kaisha
Jin NAKAMURA, Editorial Writer, *The Yomiuri Shinbun*
Seiki NISHIHIRO, Defence Adviser
Yoshihisa OHJIMI, President, CISTEC
Hiroto OHYAMA, Professor, Faculty of International Trade, Tokyo International University
Naoaki OKABE, Editorial Writer, *The Nikkei Shinbun*
Yoshikazu TAKAISHI, Managing Director, IBM Japan
Eiji UEMURA, Professor, Faculty of Law, Seikei University

INDEX